T0254170

SpringerBriefs in Applied Sciences and Technology

Nanoscience and Nanotechnology

Nanoscience and nanotechnology offer means to assemble and study superstructures, composed of nanocomponents such as nanocrystals and biomolecules, exhibiting interesting unique properties. Also, nanoscience and nanotechnology enable ways to make and explore design-based artificial structures that do not exist in nature such as metamaterials and metasurfaces. Furthermore, nanoscience and nanotechnology allow us to make and understand tightly confined quasi-zero-dimensional to two-dimensional quantum structures such as nanoplatelets and graphene with unique electronic structures. For example, today by using a biomolecular linker, one can assemble crystalline nanoparticles and nanowires into complex surfaces or composite structures with new electronic and optical properties. The unique properties of these superstructures result from the chemical composition and physical arrangement of such nanocomponents (e.g., semiconductor nanocrystals, metal nanoparticles, and biomolecules). Interactions between these elements (donor and acceptor) may further enhance such properties of the resulting hybrid superstructures. One of the important mechanisms is excitonics (enabled through energy transfer of exciton-exciton coupling) and another one is plasmonics (enabled by plasmon-exciton coupling). Also, in such nanoengineered structures, the light-material interactions at the nanoscale can be modified and enhanced, giving rise to nanophotonic effects.

These emerging topics of energy transfer, plasmonics, metastructuring and the like have now reached a level of wide-scale use and popularity that they are no longer the topics of a specialist, but now span the interests of all "end-users" of the new findings in these topics including those parties in biology, medicine, materials science and engineerings. Many technical books and reports have been published on individual topics in the specialized fields, and the existing literature have been typically written in a specialized manner for those in the field of interest (e.g., for only the physicists, only the chemists, etc.). However, currently there is no brief series available, which covers these topics in a way uniting all fields of interest including physics, chemistry, material science, biology, medicine, engineering, and the others.

The proposed new series in "Nanoscience and Nanotechnology" uniquely supports this cross-sectional platform spanning all of these fields. The proposed briefs series is intended to target a diverse readership and to serve as an important reference for both the specialized and general audience. This is not possible to achieve under the series of an engineering field (for example, electrical engineering) or under the series of a technical field (for example, physics and applied physics), which would have been very intimidating for biologists, medical doctors, materials scientists, etc.

The Briefs in NANOSCIENCE AND NANOTECHNOLOGY thus offers a great potential by itself, which will be interesting both for the specialists and the non-specialists.

More information about this series at http://www.springer.com/series/11713

Ali Rafiei Miandashti · Susil Baral
Eva Yazmin Santiago
Larousse Khosravi Khorashad
Alexander O. Govorov · Hugh H. Richardson

Photo-Thermal Spectroscopy with Plasmonic and Rare-Earth Doped (Nano)Materials

Basic Principles and Applications

Springer

Ali Rafiei Miandashti
Department of Chemistry and Biochemistry
Ohio University
Athens, OH, USA

Susil Baral
Department of Chemistry
and Chemical Biology
Cornell University
Ithaca, NY, USA

Eva Yazmin Santiago
Department of Physics and Astronomy
Ohio University
Athens, OH, USA

Larousse Khosravi Khorashad
Department of Physics and Astronomy
Ohio University
Athens, OH, USA

Alexander O. Govorov
Department of Physics and Astronomy
Ohio University
Athens, OH, USA

and

Institute of Fundamental and Frontier
Sciences
University of Electronic Science and
Technology of China
Chengdu, China

Hugh H. Richardson
Department of Chemistry and Biochemistry
Ohio University
Athens, OH, USA

ISSN 2191-530X ISSN 2191-5318 (electronic)
SpringerBriefs in Applied Sciences and Technology
ISSN 2196-1670 ISSN 2196-1689 (electronic)
Nanoscience and Nanotechnology
ISBN 978-981-13-3590-7 ISBN 978-981-13-3591-4 (eBook)
https://doi.org/10.1007/978-981-13-3591-4

Library of Congress Control Number: 2018963985

This Springer imprint is published by the registered company Springer Nature Singapore Pte Ltd.
The registered company address is: 152 Beach Road, #21-01/04 Gateway East, Singapore 189721, Singapore

Preface

The study of noble metal nanoparticles has attracted a plethora of scientific efforts due to their strong absorption and scattering of electromagnetic radiation properties. The extremely large light absorption at the plasmon resonance leads to large local temperatures that can be used for the remote release of encapsulated material, melting of strands of DNA, thermal therapy of tumors, and controlled manipulation of phase transitions of phospholipid membranes. The total amount of heat generated under optical illumination can be estimated in a simple way as the total optical absorption rate, but it is extremely challenging to measure the local temperature change. In this booklet, we aim to present recent experimental results in photothermal spectroscopy for measuring nanoscale temperature and to provide a theoretical foundation for understanding the photothermal properties of gold nanostructures and the thermal interactions with the surrounding.

In our groups, we have been working on the theoretical and experimental understanding of optical and photothermal properties of plasmonic nanoparticles for more than a decade. The main focus of researches in our groups has been ways to measure the temperature of optically illuminated plasmonic nanoparticles through luminescence thermometry of rare-earth doped nanoparticles and thin film. In the introduction, theoretical chapter, and following experimental chapters, we overview background, optical, and photothermal properties of plasmonic nanoparticles and experimental approaches undertaken to measure the photothermal properties using rare-earth doped (nano)materials.

We believe that a brief collection of recent development on thermometry of noble metal nanoparticles using rare-earth doped materials would benefit the general audience and help the scientific community to further progress the study of the photothermal phenomenon. The majority of this booklet is the collection of recent articles published in Dr. Richardson and Dr. Govorov's group over the past decade.

The figures and content are reprinted with permission from journal publishers, and credit is given at the beginning of each chapter and figures captions. Mr. Ali Rafiei Miandashti and Dr. Susil Baral have contributed equally in the preparation of the final draft of the booklet manuscript.

Athens, USA	Ali Rafiei Miandashti
Ithaca, USA	Susil Baral
Athens, USA	Eva Yazmin Santiago
Athens, USA	Larousse Khosravi Khorashad
Athens, USA, and Chengdu, China	Alexander O. Govorov
Athens, USA	Hugh H. Richardson

Contents

1 Introduction 1
References 3

2 Theory of Photo-Thermal Effects for Plasmonic Nanocrystals and Assemblies 5
Eva Yazmin Santiago, Larousse Khosravi Khorashad and Alexander O. Govorov
2.1 Introduction 5
2.2 Optical Properties of Single Nanoparticles and Nanoparticle Clusters 6
2.2.1 Mie Theory 7
2.2.2 Effective Medium Theory 9
2.2.3 Effect of Geometry of the System 10
2.2.4 Effect of Nanoparticle Material and Its Surrounding Medium 12
2.3 Optically Generated Heat Effects 13
2.3.1 Single Spherical Nanoparticles 14
2.3.2 Ensemble of Nanoparticles 17
2.3.3 Thermal Complexes with Hot Spots 18
References 22

3 Nanoscale Temperature Measurement Under Optical Illumination Using AlGaN:Er^{3+} Photoluminescence Nanothermometry 23
Susil Baral, Ali Rafiei Miandashti and Hugh H. Richardson
3.1 Introduction 23
3.2 AlGaN:Er^{3+} Photoluminescence Nanothermometry 23
3.3 Experimental Details of AlGaN:Er^{3+} Photoluminescence Nanothermometry 25
References 29

4 Comparison of Nucleation Behavior of Surrounding Water Under Optical Excitation of Single Gold Nanostructure and Colloidal Solution 31
Susil Baral, Ali Rafiei Miandashti and Hugh H. Richardson
4.1 Introduction 31
4.2 Temperature Changes and Phase Transformation with Gold Nano-wrenches 31
4.3 Dynamic Temperature Changes and Phase Transformation with Gold Nano-wrenches 32
4.4 Temperature Measurements of Optically Excited Colloidal Gold Nanoparticles 35
4.5 Temperature Measurements Probing Convection of the Liquid During Laser Excitation of a Colloidal Nanoparticle Solution . . . 35
References 38

5 Effect of Ions and Ionic Strength on Surface Plasmon Extinction Properties of Single Plasmonic Nanostructures 39
Susil Baral, Ali Rafiei Miandashti and Hugh H. Richardson
5.1 Introduction 39
5.2 Measurement of Nanoscale Temperature Change on Optically Excited Gold Nanowires Using AlGaN:Er^{3+} Nanothermometry 40
5.3 Dynamic Temperature Measurements on Single Gold Nanowire Using Flow Cell 42
5.4 Model of Heat Transfer 42
5.5 Absorption Measurements on Gold Nanoparticle(s)/ Gold Nanorod(s) 43
5.6 Absorption and Temperature Measurements on a Same Gold Nanoparticle(s) 45
5.7 Single Nanowire Dark-Field Scattering Measurements 46
5.8 Single Nanoparticle(s) Emission Measurements 47
5.9 Calculation of Absorption Cross Section of a Nanowire 47
5.10 Langmuir Model of Charge Occupancy and Effect on Absorption Attenuation 49
References 49

6 Photothermal Heating Study Using Er_2O_3 Photoluminescence Nanothermometry 51
Susil Baral, Ali Rafiei Miandashti and Hugh H. Richardson
6.1 Introduction 51
6.2 Temperature Calibration of Erbium Oxide Photoluminescence . . . 52
6.3 Temperature Profile of Single Gold Nanodot 54
6.4 Temperature Measurement Inside a Microbubble 58
6.5 Drawbacks/Limitations of the Technique 59
References 60

7 Nanoscale Temperature Study of Plasmonic Nanoparticles Using NaYF4:Yb^{3+}:Er^{3+} Upconverting Nanoparticles 63
Ali Rafiei Miandashti, Susil Baral and Hugh H. Richardson
7.1 Introduction 63
7.2 Temperature Calibration of $NaYF_4$:Yb^{3+},Er^{3+} Nanocrystals Photoluminescence 63
7.3 Characterization of $NaYF_4$:Yb^{3+},Er^{3+} Nanocrystals 65
7.4 Lifetime Study of $NaYF_4$:Yb^{3+},Er^{3+} Nanocrystals 67
References 71

8 Near Field Nanoscale Temperature Measurement Using AlGaN: Er^{3+} Film via Photoluminescence Nanothermometry 73
Susil Baral, Ali Rafiei Miandashti and Hugh H. Richardson
8.1 Introduction 73
8.2 Characterization of NSOM Tip and Nanoparticles 74
8.3 Sub Diffraction Near Field Photothermal Temperature Measurement 75
8.4 Steady State Near Field Photothermal Heat Study 79
8.4.1 Estimation of Cluster Radius (R_c) from Thermal Profile 80
8.5 Comparison Between Estimation of Cluster Radius (R_c) from Thermal Profile, AFM, and Changes on Er^{3+} Luminescence Intensity 81
8.6 Two Laser Steady State Data Collection Experiment 82
8.7 Scaling Law in Near Field Photothermal Heat Dissipation 83
References 86

Chapter 1
Introduction

There is Plenty of Room at the Bottom
Richard Feynman, Caltech, 1959

Perhaps it is the most common quote you might hear if you enter the world of nanoscience and nanotechnology. Feynman was not the first person that discovered the possibility of manipulating the materials in small scale. In ancient and medieval times, metal colloids (from the Greek, *kola*, meaning "*glue*") were used for colored glasses and dyes. Fine small metal particles fabricated during different ages of ancient times were composed of micro and nanosized particles. One of the most famous artworks is a Roman glasswork made in fifth century showing the King Lycargus. This remarkable cup housed in British Museum is a great example of application of nanotechnology in the ancient world. When illuminated from outside, the cup appears green whereas when illuminated from inside, the cup appears crimson red (Fig. 1.1). In 1990, scanning electron microscope images discovered that the specific cause of this transmission and scattering dependent colors was the presence of nanosized particles of silver, gold and copper (<100 nm) embedded in the glass matrix. After Romans, medieval craftsmen also exploited the addition of metallic particles to create stained glass windows. Unique property of nanosized metal particles did not limit to gold, silver and copper and they have expanded to other metals, oxides and semiconductors ever since.

Discovering the optical, electrical, and catalytic properties of the noble metal nanomaterials in recent years has become the basis of modern nanoscience and nanotechnology. With increasing application of nanoscale systems to drive various processes, understanding fundamental properties of metal nanomaterials at the nanoscale has become inevitable as it not only helps to expand their scope but also permits evaluating the efficiency and performance of nanosized systems [1, 2]. It is the unique properties of metal nanomaterials that make them great candidates for their applications in various aspects of science and technology.

The unique optical, electrical and catalytical properties of metal nanoparticles stem from strong absorption and scattering of electromagnetic radiation at the nanoscale [1, 3, 4]. The origin of this unique light extinction property is attributed to the phenomenon of surface plasmon resonance [1, 2, 5–8]. Light absorption by

A. R. Miandashti et al., *Photo-Thermal Spectroscopy with Plasmonic and Rare-Earth Doped (Nano)Materials*, Nanoscience and Nanotechnology,
https://doi.org/10.1007/978-981-13-3591-4_1

Fig. 1.1 Lycargus cup illuminated from outside (left) and inside (right). Image reprinted with permission from British Museum images

plasmonic nanoparticles is greatly enhanced because of surface plasmon resonance, while the emission quantum efficiency of the metal nanoparticles is very negligible. The absorbed electromagnetic radiation is, therefore, converted into the heat energy through a series of non-radiative relaxation dynamics involving electron-electron and electron–phonon interactions, resulting in local heating of the nanoparticle/nanostructure under optical illumination [8–10]. This non-radiative transformation of the light energy into the heat energy is referred to as 'photothermal process/effect'. As this heat generation is direct consequence of the surface plasmon absorption, the total amount of heat generated depends on the nature of the metal, shape, size and number of nanoparticles present in the system of interest.

The 'photothermal' heat generated by metal nanoparticles can be utilized to drive processes in numerous applications, and the heat generation and dissipation properties of plasmonic nanoparticles and its surrounding impacts numerous technologies where nanoscale heaters are used to drive processes for solar steam generation [11], thermal energy harvesting [12], photothermal therapy [13], drug delivery [14], water sterilization [11], biological imaging [15], biological actuation [16], etc. In these applications, the efficiency and performance of a nanosystem is dictated by the heat generation and or the heat dissipation from optically heated nanoheaters. Understanding the heat generation and heat dissipation properties of a nanosystem and its surrounding is, therefore, vital for its efficiency and performance.

The total amount of heat generated under optical illumination can be estimated in a simpler way as the total optical absorption rate [9], but it is extremely challenging to measure the local temperature change on nanoparticles experimentally. Measuring temperature change at a nanoscale can be used as an important parameter to understand the fundamental properties as well as applications of plasmonic materials under optical illumination. A number of techniques and approaches have been developed in recent years to measure the local temperature change in plasmonic nanostructures. Fluorescent Intensity Ratio (FIR) technique which is based on measuring the intensity of two closely spaced energy levels is one of the most reliable methods [17–19]. In this technique, nanoscale temperature and thermal profile of nanostructures under optical illumination was investigated [17]. In Richardson's lab, Carlson et al. developed this technique with thermal sensor film of $Al_{0.94}Ga_{0.06}N$ with embedded Er^{3+} ions [17]. This thermal sensor film emit photoluminescence light in the green region of electromagnetic spectrum. Also other photoluminescence emitters of Er^{3+} such as Er_2O_3 and $NaYF_4$:Er^{3+}:Yb^{3+} nanoparticles have been introduced as effective and reliable luminescent thermal sensors [20, 21]. All these optical thermal sensors work based on fluorescent ratio thermometry where the relative intensities of green energy bands are used for calculating the local temperature. Using these luminescent films/nanoparticles, optical and photothermal properties of gold nanostructures is studied.

In this booklet some theoretical background on photothermal properties of plasmonic nanoparticles/nanostructures and some experimental work probing the fundamental photothermal properties using Er^{3+} emission nanothermometry is presented. Gold nanostructures are optically excited and the nanoscale temperature change under optical illumination is measured using trivalent lanthanide ion photoluminescence nanothermometry. As the temperature change is direct consequence of the optical absorption of plasmonic nanostructures; it can, therefore be used as a fundamental parameter to study the photothermal response of a system under different surrounding and experimental conditions.

References

1. Link S, El-Sayed MA (2000) Shape and size dependence of radiative, non-radiative and photothermal properties of gold nanocrystals. Int Rev Phys Chem 19(3):409–453
2. Eustis S, El-Sayed MA (2006) Why gold nanoparticles are more precious than pretty gold: noble metal surface plasmon resonance and its enhancement of the radiative and nonradiative properties of nanocrystals of different shapes. Chem Soc Rev 35(3):209–217
3. El-Sayed MA (2001) Small is different: some interesting properties of material confined to the manometer length scale of different shapes. In: Abstracts of papers of the American chemical society, 113-CHED, vol 222
4. El-Sayed MA (2002) Shape dependent properties of some semiconductor and metallic nanocrystals. In: Abstracts of papers of the american chemical society, 021-PHYS, vol 224
5. Link S, El-Sayed MA (1999) Size and temperature dependence of the plasmon absorption of colloidal gold nanoparticles. J Phys Chem B 103(21):4212–4217
6. Maier SA, Brongersma ML, Kik PG, Meltzer S, Requicha AAG, Atwater HA (2001) Plasmonics—a route to nanoscale optical devices. Adv Mater 13(19):1501–1505
7. Jain PK, Lee KS, El-Sayed IH, El-Sayed MA (2006) Calculated absorption and scattering properties of gold nanoparticles of different size, shape, and composition: applications in biological imaging and biomedicine. J Phys Chem B 110(14):7238–7248
8. Huang X, El-Sayed MA (2010) Gold nanoparticles: optical properties and implementations in cancer diagnosis and photothermal therapy. J Adv Res 1(1):13–28

9. Govorov AO, Richardson HH (2007) Generating heat with metal nanoparticles. Nano Today 2(1):30–38
10. Sau TK, Rogach AL, Jaeckel F, Klar TA, Feldmann J (2010) Properties and applications of colloidal nonspherical noble metal nanoparticles. Adv Mater 22(16):1805–1825
11. Neumann O, Urban AS, Day J, Lal S, Nordlander P, Halas NJ (2013) Solar vapor generation enabled by nanoparticles. ACS Nano 7(1):42–49
12. Lee SW, Yang Y, Lee H-W, Ghasemi H, Kraemer D, Chen G, Cui Y (2014) An electrochemical system for efficiently harvesting low-grade heat energy. Nat Commun 5:3942
13. Huang XH, Jain PK, El-Sayed IH, El-Sayed MA (2008) Plasmonic photothermal therapy (PPTT) using gold nanoparticles. Lasers Med Sci 23(3):217–228
14. Wijaya A, Schaffer SB, Pallares IG, Hamad-Schifferli K (2009) Selective release of multiple DNA Oligonucleotides from gold nanorods. ACS Nano 3(1):80–86
15. Jiao PF, Zhou HY, Chen LX, Yan B (2011) Cancer-targeting multifunctionalized gold nanoparticles in imaging and therapy. Curr Med Chem 18(14):2086–2102
16. Lapotko DO, Lukianova-Hleb EY, Oraevsky AA (2007) Clusterization of nanoparticles during their interaction with living cells. Nanomedicine 2(2):241–253
17. Carlson MT, Khan A, Richardson HH (2011) Local temperature determination of optically excited nanoparticles and nanodots. Nano Lett 11(3):1061–1069
18. Paez G, Strojnik M (2003) Erbium-doped optical fiber fluorescence temperature sensor with enhanced sensitivity, a high signal-to-noise ratio, and a power ratio in the 520–530- and 550–560-nm bands. Appl Opt 42(16):3251–3258
19. Berthou H, Jörgensen CK (1990) Optical-fiber temperature sensor based on upconversion-excited fluorescence. Opt Lett 15(19):1100–1102
20. Debasu ML, Ananias D, Pastoriza-Santos I, Liz-Marzan LM, Rocha J, Carlos LD (2013) All-in-one optical heater-thermometer nanoplatform operative from 300 to 2000 K based on Er3 + emission and blackbody radiation. Adv Mater 25(35):4868–4874
21. Vetrone F, Naccache R, Zamarron A, Juarranz de la Fuente A, Sanz-Rodriguez F, Martinez Maestro L, Martin Rodriguez E, Jaque D, Garcia Sole J, Capobianco JA (2010) Temperature sensing using fluorescent nanothermometers. ACS Nano 4(6):3254–3258

Chapter 2
Theory of Photo-Thermal Effects for Plasmonic Nanocrystals and Assemblies

Eva Yazmin Santiago, Larousse Khosravi Khorashad and Alexander O. Govorov

2.1 Introduction

Plasmons are quasiparticles derived from a collective oscillation of electron gas. When plasmons are coupled with an electromagnetic wave, another quasiparticle named plasmon polariton is created. In this chapter, we mainly focus on optically driven plasmons and so we shall refer to plasmon polaritons as plasmons. Additionally, the plasmon resonance frequency is the frequency at which the electrons in the material would oscillate upon electromagnetic illumination. Therefore, if the light frequency matches the plasmon resonance frequency, the oscillation amplitude reaches its maximum. Due to conservation of energy, this leads to a maximum light absorption which strongly enhances the heat generation in a system. In particular, plasmonic nanocrystals can efficiently generate heat under light stimulation.

Metal nanoparticles (NPs) are more efficient due to their high electron density which implies high absorption and heat generation. It is also apparent from bulk optical properties of metal particles that metals, contrary to dielectrics, comprise the imaginary part of the permittivity, which accounts for the high resistive losses. Two of the most popular metals are gold (Au) and silver (Ag) because their plasmonic resonances fall in the visible spectrum of electromagnetic radiation. In addition, Au nanoparticles are more stable chemically and physically. The photothermal heat generated in a plasmonic system is theoretically described by Maxwell's equations and heat transfer theory. Maxwell's equations produce the spatial distribution of the heating intensity in the system, which is stronger in the metal component. However, controlling this heating effect in terms of efficiency and localization can present a difficult challenge.

This chapter gives a review of the theory that describes the photothermal properties of a plasmon nanostructure. In Sect. 2.2, we describe the optical response

A. R. Miandashti et al., *Photo-Thermal Spectroscopy with Plasmonic and Rare-Earth Doped (Nano)Materials*, Nanoscience and Nanotechnology,
https://doi.org/10.1007/978-981-13-3591-4_2

of single NPs and clusters/aggregates composed of NPs. In particular, we focus on spherical NPs surrounded by a homogenous medium. We introduce Mie theory, the quasistatic approximation and effective medium theories to explain the interaction between light and matter. We also explain the influence of the morphology of system on such interaction. In Sect. 2.3, we use the theory developed in Sect. 2.2 to describe the heat generation and transfer in a plasmonic complex. We, once again, focus on individual NPs and clusters. Additionally, the inter-particle interaction that occurs in dimers and trimer will be discussed.

2.2 Optical Properties of Single Nanoparticles and Nanoparticle Clusters

We can theoretically describe the heat generation by a single NP, which is useful in diluted systems with strong light intensities. The heat generated in the NP is diffused to the surrounding medium, leading to an increase of the temperature at its surface. Light extinction is the combination of the scattering and absorption produced by its interaction with the nanocrystal. For small NPs, where their size is much smaller than the incident wavelength, the absorption dominates over the scattering. As the size increases, the scattering becomes predominant, which is displayed in Fig. 2.1. We also observe that a red-shift of the peaks occurs, which is a convenient trait since it allows the system to be tuned.

To quantify this phenomenon, we introduce the absorption, scattering and extinction cross sections, denoted as σ_{abs}, σ_{scat} and σ_{ext} respectively, with units of area. Physically, the extinction cross section is the virtual cross section of a perfectly opaque particle that absorbs and scatters the same amount of radiation as the NP in our system.

To describe the photothermal process in a plasmon structure, we focus on the absorption cross section, defined as

$$\sigma_{abs} = \frac{P_{abs}}{I_0},$$

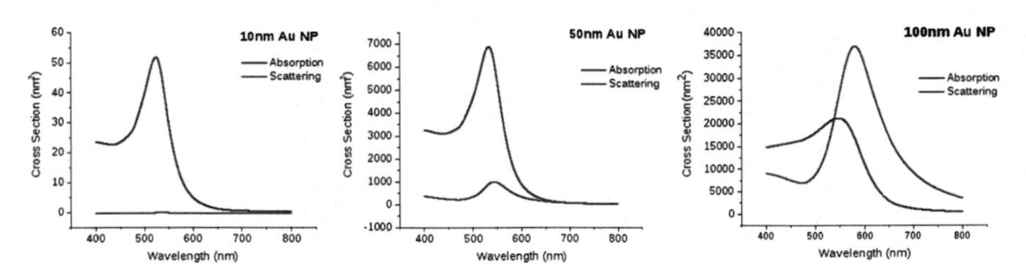

Fig. 2.1 Absorption and scattering cross sections of AuNPs submerged in water with different sizes

where P_{abs} is the absorption power and I_0 is the incident flux. The absorption cross section allows the calculation of the molar absorption coefficient (ε_{abs}), which is more common in biological applications. This quantity is proportional to σ_{abs}, in the form

$$\varepsilon_{abs} = \left(\frac{N_A}{0.23}\right) \times 10^{-4} \cdot \sigma_{abs},$$

where N_A is Avogadro's number and σ_{abs} has the unit of cm^2. The molar absorption then has the unit of $M^{-1}cm^{-1}$. The same proportionality is preserved between the molar scattering and extinction coefficients (ε_{scat}, ε_{ext}) and the scattering and extinction cross sections (σ_{scat}, σ_{ext}) such that

$$\varepsilon_{scat} = \left(\frac{N_A}{0.23}\right) \times 10^{-4} \cdot \sigma_{scat},$$

$$\varepsilon_{ext} = \left(\frac{N_A}{0.23}\right) \times 10^{-4} \cdot \sigma_{ext}.$$

An additional photothermal effect mechanism is the collective heat generation in a system with more than one particle. This is important for NP clusters, where the final result can be considered as an accumulation of heat produced by all the NPs involved, where each individual contribution is relatively small.

The following two subsections introduce the theory that allows the calculation of the absorption, scattering and extinction cross sections for a plasmonic nanostructure. To obtain these optical properties we consider a system described by an individual NP, a cluster or an aggregate with a dielectric constant ε_{NP} surrounded by a dielectric medium (ε_{matrix}). We should note that although we call ε a constant, it depends on the frequency ω and consequently, the wavelength λ of the incident electromagnetic wave. The constant aspect of this quantity is due to the assumption that the materials used in this study are spatially homogenous.

For a simple system like a spherical or an ellipsoidal NP, an exact solution of the Maxwell equations is provided by Mie theory (Sect. 2.2.1). When the system is small enough, we can simplify the solution through a quasistatic approximation (Sect. 2.2.1.1). To theoretically describe the light-matter interaction of a cluster or an aggregate, we first employ an effective medium theory (EMT), which is described in Sect. 2.2.2.

2.2.1 *Mie Theory*

To obtain the exact solution of the optical response regarding a sphere of arbitrary size embedded in a homogeneous medium we rely on Mie Theory, developed by Gustav Mie in 1908. Other exact solutions, such as light scattering from spheroids and infinite cylinders, also exist [1]. However, obtaining a solution for Maxwell's equations for the light-matter interaction for systems with different geometries is laborious and most of the time requires a numerical solution.

According to Mie theory, the extinction, scattering and absorption cross sections of a sphere of radius R_{NP} in a matrix are given as

$$\begin{aligned}\sigma_{ext} &= \pi R_{NP}^2 \times \frac{2}{x^2}\sum_{n=1}^{\infty}(2n+1)Re[a_n + b_n],\\ \sigma_{scat} &= \pi R_{NP}^2 \times \frac{2}{x^2}\sum_{n=1}^{\infty}(2n+1)\left(|a_n|^2 + |b_n|^2\right),\\ \sigma_{abs} &= \sigma_{ext} - \sigma_{scat},\end{aligned} \tag{2.1}$$

where a_n and b_n are the Mie coefficients and are expressed as

$$\begin{aligned}a_n &= \frac{\psi_n'(y)\psi_n(x) - \sqrt{\varepsilon_{NP}/\varepsilon_{matrix}}\psi_n(y)\psi_n'(x)}{\psi_n'(y)\zeta_n(x) - \sqrt{\varepsilon_{NP}/\varepsilon_{matrix}}\psi_n(y)\zeta_n'(x)},\\ b_n &= \frac{\sqrt{\varepsilon_{NP}/\varepsilon_{matrix}}\psi_n'(y)\psi_n(x) - \psi_n(y)\psi_n'(x)}{\sqrt{\varepsilon_{NP}/\varepsilon_{matrix}}\psi_n'(y)\zeta_n(x) - \psi_n(y)\zeta_n'(x)},\end{aligned}$$

where $x = \sqrt{\varepsilon_{matrix}}k_{vac}R_{NP}$ is the size parameter, and $y = \sqrt{\varepsilon_{NP}}k_{vac}R_{NP}$. These equations include the Riccati-Bessel functions $\psi_n(x)$ and $\zeta_n(x)$ and their derivatives, as well as the wave vector of light in vacuum, $k_{vac} = \frac{2\pi}{\lambda} = \frac{\omega}{c}$. We should note that the sphere considered in this calculation can be a single nanoparticle of any size or a spherical cluster composed of an ensemble of nanoparticles. In Sect. 2.2.2, we will learn how to manipulate the dielectric constant of the latter to proceed with the equations above.

2.2.1.1 Quasistatic Approximation

When the size of the NP or cluster is much smaller than the incident light wavelength, the external electric field near the NP can be considered spatially uniform. The dipole plasmon resonance is then dominant over higher modes of the plasmon excitation, as it is shown in Fig. 2.2a. Therefore, in the Mie theory solution we only consider the first term ($n = 1$). In the case of a spherical NP [2].

$$\sigma_{abs} = \pi R_{NP}^2 \times 4xIm\left\{\frac{\varepsilon_{NP} - \varepsilon_{matrix}}{\varepsilon_{NP} + 2\varepsilon_{matrix}}\right\},$$

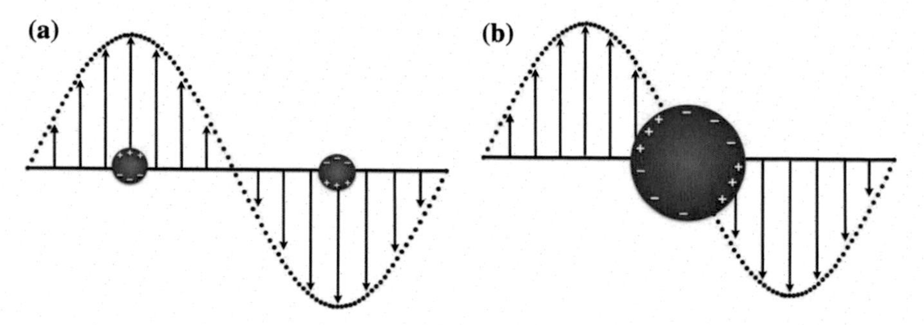

Fig. 2.2 Charge distribution in a spherical NP when its size is: **a** smaller or **b** comparable to the electromagnetic wavelength

which is sometimes expressed as

$$\sigma_{abs} = \pi R_{NP}^2 \times \frac{4R_{NP}}{3\sqrt{\varepsilon_{matrix}}}\frac{\omega}{c_o}\left|\frac{3\varepsilon_{matrix}}{\varepsilon_{NP} + 2\varepsilon_{matrix}}\right|^2 Im\{\varepsilon_{NP}\},$$

where ω and c_o are the angular frequency and the speed of light, respectively.

Having σ_{scat} with the same aproach for σ_{abs}, the extinction cross section is obtained from Eq. (2.1) [2]. As the size of the NP or cluster increases, this approximation becomes less effective and we must consider more terms in the Mie Theory solution. Figure 2.2b displays a sphere with a size comparable to the incident wavelength for which we must consider the dipole and multipole terms.

2.2.2 *Effective Medium Theory*

Photothermal systems are often described by NP clusters or aggregates. However, the interaction between the NPs greatly complicates the exact calculation of the optical and photothermal properties of the system. Therefore, an effective medium theory can be used to simplify the problem. There are several approximations developed to find an effective dielectric constant of a spherical cluster; one of the most well-known is the Maxwell-Garnett equation [3], expressed as

$$\varepsilon_{eff} = \varepsilon_v \frac{\varepsilon_{metal}(1 + 2f) - \varepsilon_v(2f - 2)}{\varepsilon_{metal}(1 - f) + \varepsilon_v(2 + f)},$$

where ε_v is the dielectric constant of the voids and f is the filling (fraction) factor, which is the ratio of the volumes of the metallic component and the void. Similarly, the Bruggeman approximation states that

$$\varepsilon_{eff} = \frac{1}{4}(\beta \pm \sqrt{\beta^2 + 8\varepsilon_{metal}\varepsilon_v}),$$
$$\beta = (3f - 1)\varepsilon_{metal} + (3f - 1)\varepsilon_v,$$

where the sign $\pm$ is chosen so that $Im\{\varepsilon_{eff}\} > 0$. In principle, an effective medium theory consists of a calculation of a weighted average of the components in the system by using the filling factor. Although it fails to consider any inter-particle interaction, it can be considered as an acceptable first approximation to a more complex problem, especially for a diluted aggregate.

Once an effective dielectric constant is obtained for a cluster or aggregate, we then apply Mie Theory to obtain the overall response of the system by considering that the cluster is now a homogeneous sphere with a dielectric constant ε_{eff}. Due to a difference in the filling fraction, clusters respond differently to light than NPs ($f = 1$) of the same size. Large Au clusters, in particular, have a stronger absorption in the red and near-infrared wavelength intervals while individual Au spheres primarily scatter the incident light. The physical explanation is that a cluster is more heterogeneous, allowing light to penetrate it much better and therefore, its absorption of light is more efficient.

2.2.3 *Effect of Geometry of the System*

As mentioned in Sect. 2.2.1, obtaining the optical response of an arbitrary system can be very rigorous and, in most cases, is impossible. Therefore, it is convenient to find an approximation. One alternative is the use of a computational method or software. Some efficient numerical methods are: the discrete dipole approximation (DDA), the finite element method, the finite difference method, and spectral representation.

The variation of the size and shape of a metal NP modifies its optical response [5] because the incident electromagnetic wave causes the charges inside the NP to rearrange differently. This affects the plasmon resonance frequency, which means that we can have some control over the photothermal process in the system through the manipulation of its geometry. This is of great interest since technological advances allow us to synthesize particles with a wide variety of shapes. In Fig. 2.3, we can observe the shift of the resonance peak for different NP geometries for a particular incident electromagnetic wave.

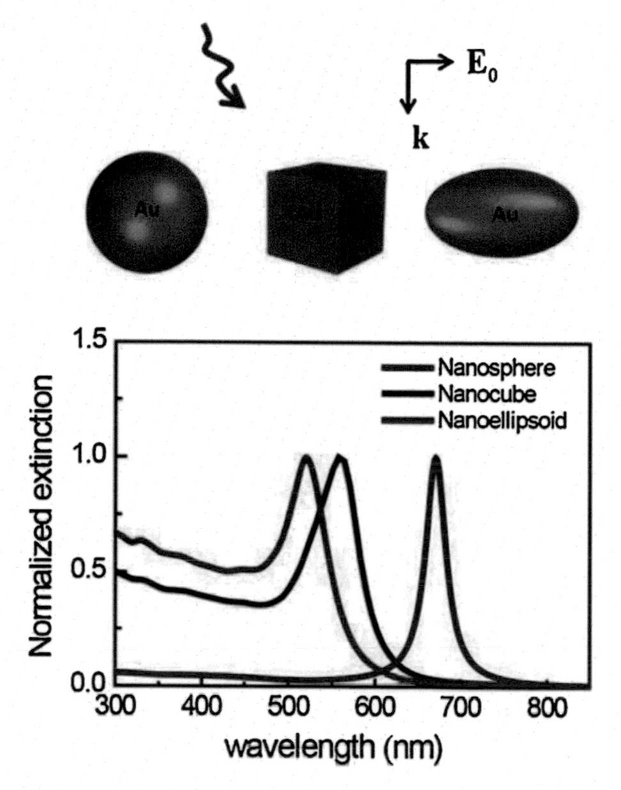

Fig. 2.3 The normalized extinction for particles with different shapes: sphere, cube, ellipsoid. Reprinted from Ref. [4] with permission from Elsevier.

In the case of a metal spheroid in a homogeneous medium, the light-matter interaction shows three plasmon resonances, each corresponding to an axis in the coordinate system. This is due to the different oscillation of the electrons along each of its axes. Therefore, by varying the length of each axis of the NP we can tune its optical response [1, 6]. The dipolar polarizability for an ellipsoid is given by

$$\alpha_i(\omega) = \frac{V_e}{4\pi} \frac{\varepsilon_{NP} - \varepsilon_{matrix}}{\varepsilon_{matrix} + L_i(\varepsilon_{NP} - \varepsilon_{matrix})},$$

where V_e is the volume of the ellipsoid and L_i is the depolarization factor in the direction i, which depends on the geometry of the NP. This modifies the optical response of the system and therefore, changes its thermal properties as well.

In many cases, we deal with multiple particles that interact with each other, exhibiting interesting photothermal properties. In Fig. 2.4 we observe the production of multiple resonance peaks for a trimer [7]. We will study these types of structures later in the chapter.

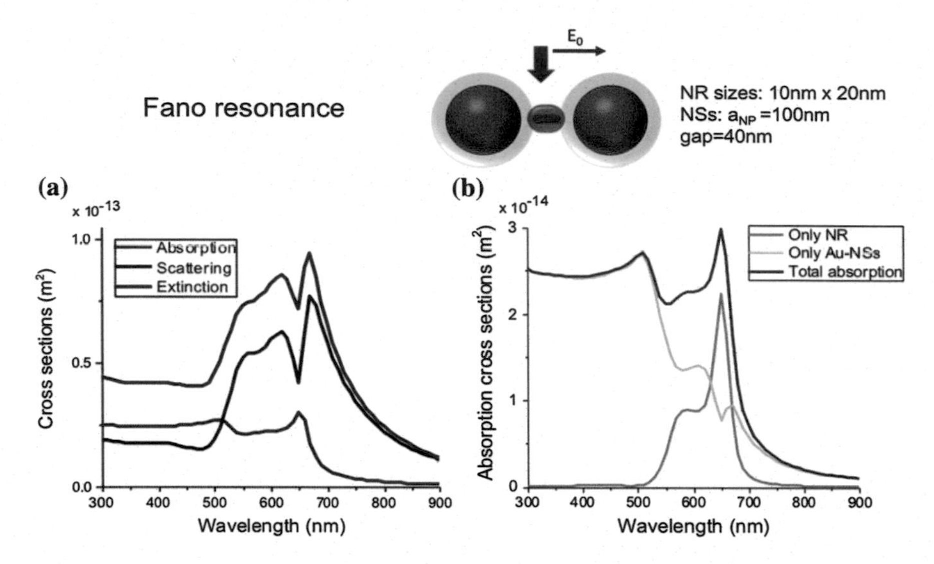

Fig. 2.4 **a** Absorption, scattering and extinction cross sections for a NS–NR–NS system. **b** Absorption cross section of the components in the same system. Reprinted (adapted) with permission from Ref. [7] (Copyright 2013 American Chemical Society)

2.2.4 Effect of Nanoparticle Material and Its Surrounding Medium

Since the response of a system when interacting with light depends on the frequency, the NP composition and its surrounding medium, or matrix, have an influence on such interaction. This is because the dielectric function ε is frequency dependent. This parameter depends on the material of each component of the system and for anisotropic NP, it also varies with the direction and polarization of the incident light. In this text we will only deal with isotropic systems.

In general, the dielectric constant is a complex parameter. The imaginary part is associated with the light absorption while the real part helps describes the degree of the metallic response of the NP. When $\varepsilon_{NP} < 0$, the material presents a metallic behavior while a dielectric constant where $\varepsilon_{NP} > 0$, causes the NP to exhibit a dielectric response. To do theoretical studies, in most cases, we rely on a model to obtain the dielectric function of many materials. Two well-known models are Drude and Lorentz. We also make use of experimental bulk permittivity such as Johnson and Christy [8] and Palik [9].

The influence of the surrounding medium on the optical response of the system is therefore related to its dielectric constant (ε_{matrix}). This is because it determines the configuration of the electric field near the NP, which implies a modification of the cross sections, as we noted in Eq. (2.1). Another effect corresponds to the polarization induced in the matrix by the NP; a larger ε_{matrix} induces a larger polarization charge. This modifies the charge distribution on the NP surface, causing a shift of the plasmon resonance wavelength. In particular, a NP in a homogeneous medium with a larger dielectric constant produces a resonance to shift to larger wavelengths [5].

2.3 Optically Generated Heat Effects

As it was mentioned before, optically stimulated NP generates heat that gets released to its surrounding medium. This mechanism is described by the heat transfer equation which is dependent on time t and position **r**. In the absence of a phase transformation this differential equation is given by

$$\rho(\mathbf{r})c(\mathbf{r})\frac{\partial T(\mathbf{r},t)}{\partial t} = \nabla(k(\mathbf{r})\nabla T(\mathbf{r},t)) + Q(\mathbf{r},t), \tag{2.2}$$

where $T(\mathbf{r},t)$ is the local temperature, and the material's thermal parameters $k(\mathbf{r})$, $\rho(\mathbf{r})$ and $c(\mathbf{r})$ are the thermal conductivity, mass density and specific heat, respectively. The function $Q(\mathbf{r},t)$ is local heat source and it is expressed as

$$Q(\mathbf{r},t) = \langle \mathbf{j}(\mathbf{r},t)\cdot\mathbf{E}(\mathbf{r},t)\rangle_t.$$

This equation describes the energy generated in a NP through light absorption. The functions $\mathbf{j}(\mathbf{r},t)$ and $\mathbf{E}(\mathbf{r},t)$ are the current density and the electric field induced in the system, respectively. We assume that the system is illuminated by a monochromatic external field, so that $\mathbf{E}(\mathbf{r},t) = \mathbf{E}_0(t)\cdot Re\left[e^{-i\omega t + i\mathbf{k}\cdot\mathbf{r}}\right]$, where $\mathbf{E}_0(t)$ is the real amplitude of the electric field. This electric field is obtained by solving the Maxwell equations and considering the boundary conditions in the system. This is the same method that is used to obtain the absorption cross section, as it was mentioned in previous sections. In fact, the relation between σ_{abs} and $Q(\mathbf{r},t)$ is

$$\sigma_{abs} = \frac{P_{abs}}{I_0} = \frac{1}{I_0}\int Q dV, \tag{2.3}$$

where I_0 is the light intensity inside the matrix, given by

$$I_0 = \frac{c|\mathbf{E_0}|^2\sqrt{\varepsilon_{\mathbf{matrix}}}}{8\pi}.$$

In the next sections, we describe the photothermal process in a system that is composed of a single nanoparticle or cluster in a homogenous medium.

2.3.1 Single Spherical Nanoparticles

The temperature on the outside of a spherical NP in the steady state regime ($t \rightarrow \infty$) is obtained by solving the heat transfer equation [10–13]. This distribution is given by

$$T(\mathbf{r}) = T_0 + \Delta T(\mathbf{r}), \tag{2.4}$$

where T_0 is the background temperature of the system and $\Delta T(\mathbf{r})$ is the temperature increase, expressed as

$$\Delta T(\mathbf{r}) = \frac{V_{NP} Q}{4\pi k_0 r}. \tag{2.5}$$

This equation is only valid for $r > R_{NP}$. On the other hand, inside the NP, the temperature increase is

$$\Delta T(\mathbf{r}) = \Delta T_{surface} + \Delta T_{surface} \frac{k_0}{2k_{NP}} \left(1 - \frac{r^2}{R_{NP}^2}\right), \tag{2.6}$$

where k_0 and k_{NP} are the thermal conductivity of the matrix and the spherical particle, respectively. Additionally, $\Delta T_{surface}$ is the temperature increase at $r = R_{NP}$, which is given from Eq. (2.5). Figure 2.5 shows these parameters and the temperature increase spatial distribution. We should note that although Eq. (2.6) describes a non-uniform temperature inside the particle, the fact that $k_0 \ll k_{NP}$ for the case of a AuNP in water causes the temperature to be regarded as a constant.

The local heat source (Q) for a small NP is easily obtained because the electric field induced inside the NP is uniform when it is optically stimulated. This leads to

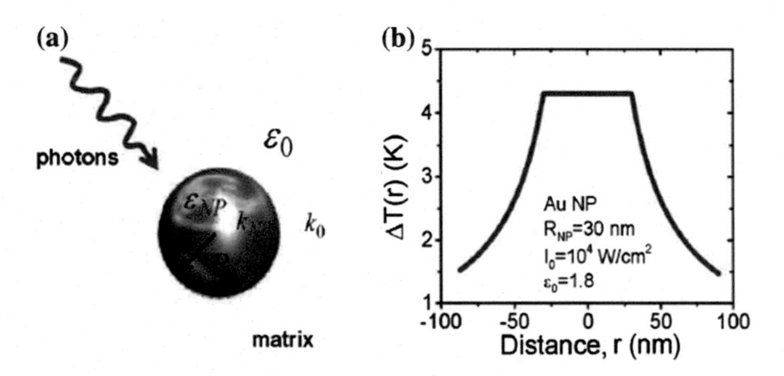

Fig. 2.5 **a** The parameters involved in a system composed of a Au spherical NP in water. **b** Temperature increase distribution of the system in A with fixed parameters. Reprinted from Ref. [12] with permission from Elsevier

a uniform absorption of light in the system which simplifies Eq. (2.3). The energy source in the quasistatic approximation is then given by

$$Q = \frac{\sigma_{abs} I_0}{V_{NP}} = \frac{\omega}{8\pi} E_0^2 \left| \frac{3\varepsilon_{matrix}}{\varepsilon_{NP} + 2\varepsilon_{matrix}} \right|^2 \mathrm{Im}\{\ \varepsilon_{NP}\}\ , \tag{2.7}$$

where E_0^2 is the amplitude of the incident radiation. From Eq. (2.5) we observe that the maximum temperature increase occurs at the surface of the nanoparticle. Therefore,

$$\Delta T_{max}(I_0) = \frac{R_{NP}^2}{3k_0} Re\left[i\omega \frac{1 - \varepsilon(\mathbf{r})}{8\pi} \left| \frac{3\varepsilon_{matrix}}{\varepsilon_{NP} + 2\varepsilon_{matrix}} \right|^2 \right] \frac{8\pi I_0}{c\sqrt{\varepsilon_{matrix}}}.$$

Its maximum dependence on the size of the nanoparticle is then

$$\Delta T_{max} \propto R_{NP}^2. \tag{2.8}$$

This proportionality is a consequence of the total rate of heat generation in the NP and the heat transfer through its surface. In fact, Eqs. (2.4)–(2.6) are also valid for small clusters. In this case, in Eq. (2.7) the dielectric constant ε_{NP} is replaced by $\varepsilon_{cluster}$ which is obtained from an EMT. In particular, from the Maxwell-Garnett equation, small clusters with a dielectric function similar to that of the matrix ($\varepsilon_{cluster} \approx \varepsilon_{matrix}$) possess an absorption cross section that increases linearly with the filling fraction (*f*) [13]. However, the shape of the spectrum does not change significantly, meaning that the resonance peak does not shift. This, of course, has a direct effect on Eq. (2.7) and the temperature distribution inside and around the cluster.

In Fig. 2.6 we can observe the dependence of the maximum temperature increase (ΔT_{max}) on the incident light wavelength for fixed parameters like the size of the sphere, the incident radiation intensity and the dielectric constants.

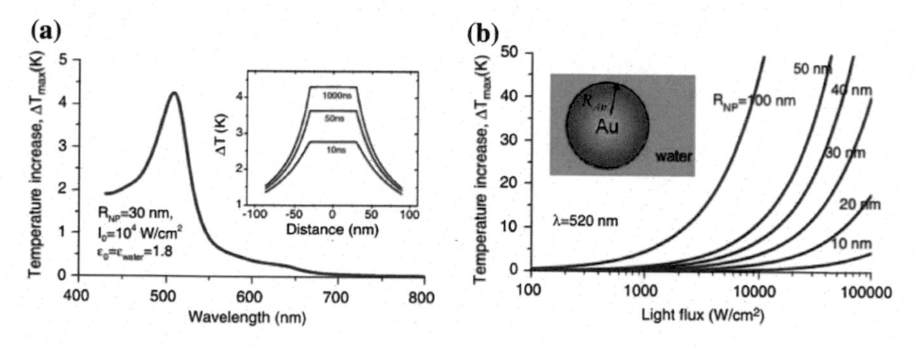

Fig. 2.6 The temperature increase as a function of: **a** the incident electromagnetic wavelength and distance from the center of the sphere (inset), **b** the light flux and distance from the center of the sphere, inset: the geometry and parameters of the AuNP. Reprinted from Ref. [14] with permission from Springer

To study large clusters, we apply an EMT and Mie theory to obtain its exact absorption spectrum. Under the steady state regime, the function Q is uniform; otherwise, we should consider the electromagnetic shadow effect [15]. However, for relatively fast heat transfer inside the cluster this effect can be neglected. Then the temperature on the surface of the cluster can be approximated to

$$\Delta T(\mathbf{r}) = \frac{V_{cluster} Q}{4\pi k_0 R_{cluster}},$$

where $R_{cluster}$ and $V_{cluster}$ are the radius and volume of the cluster, respectively. Another interesting result is that, unlike a small sphere or cluster, the size dependence of the maximum temperature increase of a large cluster is given by

$$\Delta T_{surface,cluster} \propto R_{cluster}.$$

We note the difference between this relation and the one described in Eq. (2.8). This is because for a small cluster we have that $\sigma \propto R^3_{cluster}$. However, for large sphere the absorption cross section behaves as $\sigma \propto R^2_{cluster}$ [13].

2.3.1.1 Phase Transformations

The increase of temperature at the surface of a nanoparticle and therefore, the transfer of heat to its surrounding matrix can cause it to undergo a phase transformation. The heat transfer equation that describes this process is described by Eq. (2.7), where $\frac{\partial T(\mathbf{r},t)}{\partial t} = 0$ when considering steady-state conditions.

Once again, we consider the simple case of a spherical NP in a matrix. When the NP transfers heat to its surroundings, it creates a liquid spherical shell. A schematic of a AuNP in ice is displayed in Fig. 2.7a while Fig. 2.7b includes a polymer shell.

To calculate the temperature distribution around the NP there exists a critical light power at which the matrix melts, producing the liquid shell. This critical power is given by the equations

$$\frac{Q R^2_{Au}}{3(T_{trans} - T_0)} > k_{solid},$$

$$R_b = \frac{Q R^3_{Au}}{3 k_{solid}(T_{trans} - T_0)},$$

where the local heat source Q is given by Eq. (2.7) and k_{solid} is the thermal conductivity of the solid matrix. Additionally, T_{trans} and T_0 are its phase transition temperature and the equilibrium temperature, respectively. Experimentally, the melting process is registered by observing the Raman signal intensity of the liquid/solid [16].

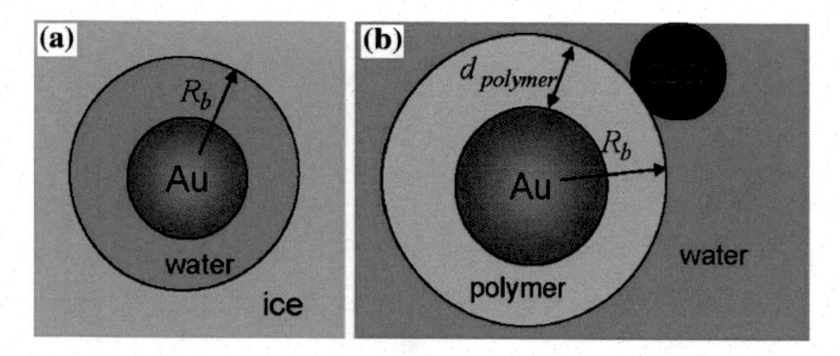

Fig. 2.7 Diagram of two systems: **a** A water shell generated around a AuNP after transferring heat to the ice surrounding it. **b** A CdTe NP attached to a polymer shell around a AuNP to detect the heat transfer. Reprinted from Ref. [12] with permission from Elsevier [12]

In the case of a polymer shell, a light-emitting CdTeNP attached to the shell surface can be used to detect the melting process [12]. This is done by observing the emission intensity of the CdTeNP, which is stimulated by the AuNP because of a resonant plasmon/exciton interaction. This process depends on the distance between both particles which increases when the polymer melts. Therefore, the melting process of a polymer shell is detected by observing a decrease of the CdTeNP emission intensity.

2.3.2 *Ensemble of Nanoparticles*

We now study the photothermal effect generated by a collection of nanoparticles. When optically excited, each NP generates heat which accumulate to strongly enhance the total effect. This mechanism is once again described by the heat transfer equation, where the local heat source is an addition of the heat generated by each individual NP. This is given by

$$Q(\mathbf{r}, t) = \sum_n Q_n(\mathbf{r}, t) = \sum_n q_n(t)\delta(\mathbf{r} - \mathbf{r_n}), \tag{2.9}$$

where $Q_n(\mathbf{r}, t)$ is the local heat source of each NP and q_n is the heat generated by the nth particle. Therefore, the heating effect is amplified by incorporating more NPs in the system, corresponding to a larger temperature increase. We should note that inter-particle interaction is then neglected in this description. For a NP system of size A and thermal diffusivity K_{matrix}, its thermal state approaches the steady-state solution of the heat transfer equation when $t \gg A/K_{matrix}$. The temperature can be approximated by

$$\Delta T_{tot}(\mathbf{r}) = \sum_n \frac{q_n}{4\pi k_0} \frac{1}{|\mathbf{r} - \mathbf{r_n}|} \approx \frac{q_0}{4\pi k_0} \int_V \frac{D(\mathbf{r})}{|\mathbf{r} - \mathbf{r}'|} d^3\mathbf{r}',$$

where $D(\mathbf{r})$ is the spatial density of identical NPs with a local power source q_0. In the limit $N_{NP}^{A/m} \gg 1$, where N_{NP} is the number of NPs in the ensemble and m is the dimensionality, the temperature inside is approximated as

$$\Delta T_{tot}(\mathbf{r}) \approx \Delta T_{max,0} \frac{R_{Au}}{\Delta} N_{NP}^{\frac{m-1}{m}}$$

for $m = 2$ and $m = 3$. For the case where $m = 1$, the estimation is given by

$$\Delta T_{tot}(\mathbf{r}) \approx \Delta T_{max,0} \frac{R_{Au}}{\Delta} \ln[N_{NP}],$$

where Δ is the average distance between NPs. Once again, we observe that the temperature raises as the number of NPs increases.

Another mechanism of interaction between the NPs that relates to the heating effect is the Coulomb interaction. The overall generated heat depends on the inter-particle interaction as a result of the plasmon-enhanced electric fields as well as the NP arrangement. This complicates the calculation since it is not as direct as the previously described mechanism and it can even lead to a decrease of the total heat dissipation even when more NPs are included. The partial screening of the electric fields inside the NPs causes the total heat produced by a NP dimer to differ from the heat generated by two individual NPs [17]. The interparticle interaction can also cause a plasmon resonance shift which, as we studied in previous sections, results in an alteration of the optical response of the system and therefore, the heat generated. In the case of the dimer, the photothermal effect depends on the light polarization because it is responsible for the nature of the plasmon-enhanced electric fields. For a pair of NPs that are randomly oriented in the matrix one can calculate an average heat generation.

2.3.3 *Thermal Complexes with Hot Spots*

When we seek to localize the temperature in a small volume, the high diffusive nature of the heat transfer becomes problematic. However, we are able to deal with it using electrodynamic hot spots and interference effects [7, 18]. We should note that so far, we have been studying the photothermal process in plasmonic systems in the regime of continuous-wave (CW) excitation. We will later see the properties of the pulsed-excitation regime and its differences with CW illumination.

In the previous section, we studied collective heating (Eq. 2.9), which is an appropriate description for large areas. To strongly localize the temperature in a confined volume, though, we turn to the local regime of heating. This regime also depends on the composition and architecture of the nanostructure, where specific conditions lead to the formation of thermal hot spots in plasmonic complexes.

To characterize the degree of localization of the temperature in a plasmonic structure we define the localization length of excess temperature, given by

$$\alpha_{\text{localization length}} = \frac{\Delta L_{\text{heating}}}{L_{\text{heater}}}, \tag{2.10}$$

where $\Delta L_{\text{heating}}$ is the dimension of the heated area in the temperature profile at $\Delta T_{\max}/2$, and L_{heater} is the size of the plasmonic heater. Additionally, we introduce the parameter

$$\alpha_{\text{temp gradient}} = \frac{\left|\frac{dT(\mathbf{r})}{dl}\right|_{\max}}{I_{\text{flux}}}, \tag{2.11}$$

where $|dT(\mathbf{r})/dl|_{\max}$ is the maximum temperature gradient that the system reaches and I_{flux} is the light flux. We note that these two figures of merit depend on the thermal and electromagnetic properties of the complex as well as its morphology. Two additional parameters are the relative efficiencies of temperature increase generation, given as

$$\text{Eff}_{\text{temp-flux}} = \frac{\Delta T_{\max}}{I_{\text{flux}}} \tag{2.12}$$

and

$$\text{Eff}_{\text{temp-abs}} = \frac{\Delta T_{\max}}{Q_{\text{abs,tot}}}, \tag{2.13}$$

where $Q_{\text{abs,tot}}$ is the rate of total absorption in the system. Different structures have been proposed to generate a thermal hot spot [7, 18]. In this text we study a NS-NR-NS complex, like the one shown in the insets of Figs. 2.8 and 2.9. This structure is composed of two large nanospheres (NS) that serve as a nano-optical antenna, and a small nanorod (NR) that acts as a nanoheater. It was found that the maximum temperature increase is reached on the NR surface, which complies with Eq. 2.5 at $r = R_{NP} = R_{NR}$.

In Figs. 2.8 and 2.9, we can observe the generation of a large temperature increase in this type of trimer, compared to an individual NS dimer and NR [7]. Figure 2.8 in particular, presents the parameter P, which is the field enhancement factor. This factor is expressed as

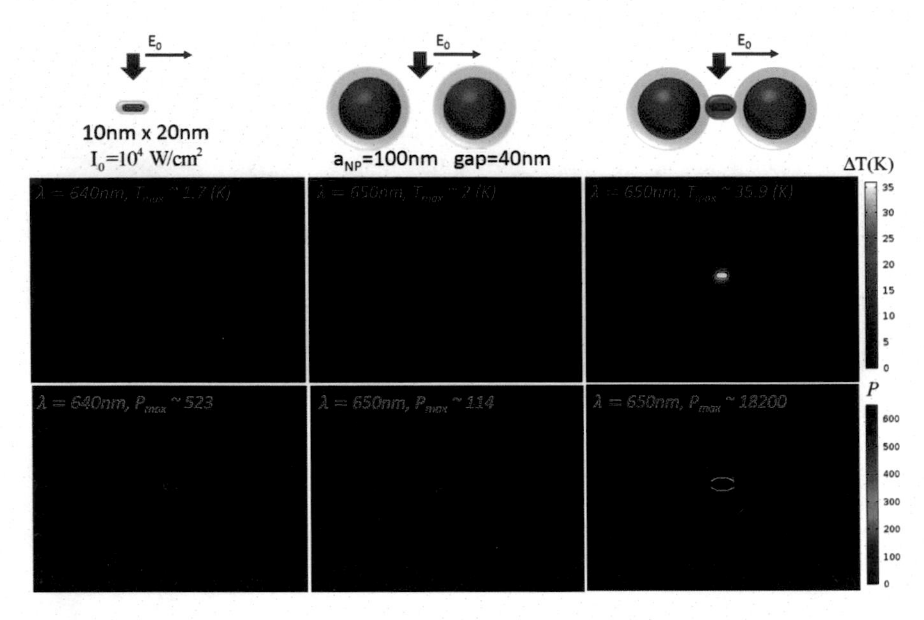

Fig. 2.8 Spatial distribution of the increase in temperature (top panel) and the corresponding field enhancement factor (bottom panel) for different systems: NR, NS–NS, NS–NR–NS. Reprinted with permission from Ref. [8] (Copyright 2013 American Chemical Society)

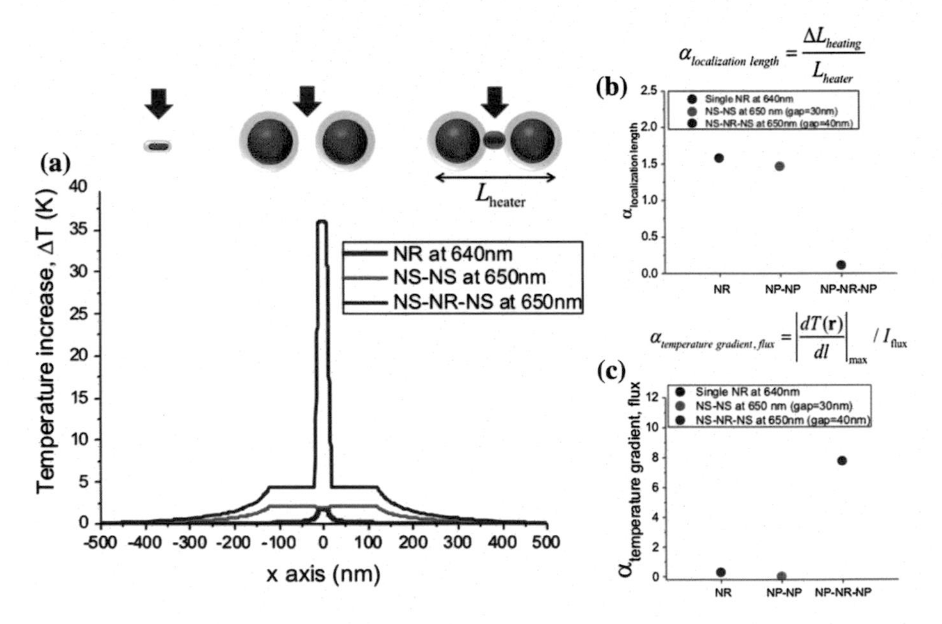

Fig. 2.9 **a** The temperature increases for different systems: NR, NS–NS, NS–NR–NS. **b** The localization length and **c** The temperature gradient of the temperature increases for the three systems. Reprinted (adapted) with permission from Ref. [7] (Copyright 2013 American Chemical Society)

$$P(\mathbf{r}) = \frac{\mathbf{E}_\omega \cdot \mathbf{E}_\omega^*}{E_0^2},$$

where $\mathbf{E}_\omega$ is the scattered electric field. We observe that the maximum field enhancement occurs in the volume that confines the NR in the trimer, compared to the single NR and the NS dimer. Furthermore, Fig. 2.9 presents $\alpha_{\text{localization length}}$ and $\alpha_{\text{temp gradient}}$ (Eqs. 2.10 and 2.11) for the same three systems.

There is a considerable difference between these parameters corresponding to the NS–NR–NS complex and the other two structures. According to Eq. (2.10), the localization length of excess temperature becomes smaller as the temperature increase becomes more localized, which is compatible with Fig. 2.9b (red dot) for the trimer. The outcome in Fig. 2.9c further supports the generation of a hot spot in this system, where the temperature gradient is much higher for the trimer (red dot) than it is for the individual NR and the NS dimer.

The production of the hot spot at the center of the trimer is due to the NR that acts as a bridge that transfer energy coherently between the two outer NSs. Studies have also been performed for an assembly composed of three NSs, although they do not exhibit the same efficiency as the NS–NR–NS system [7]. Further calculations of the efficiency parameters of Eqs. (2.12) and (2.13) show that the system of NS–NR–NS is an efficient heat generator suitable for energy applications. The reason for this behavior is because the size and geometry of the system of NS–NR–NS provide Fano resonance, which is the interaction of the broadband absorption of the nanoantenna with the sharp plasmon resonance of the NR (Fig. 2.4b).

We now focus on the pulsed regime and its comparison to the CW regime. For short times, the solution for the heat transfer equation gives

$$\Delta T(\mathbf{r}, t) = \frac{e_{tot}}{\rho_\omega c_\omega} \left(\frac{1}{2\sqrt{\pi K_{\text{diff},\omega} t}} \right)^3 e^{-r^2/4K_{\text{diff},\omega} t}.$$

Also, the figures of merit given previously become dynamic parameters, which are approximated as [7]

$$\alpha^{\text{dynamic}}_{\text{localization length}} \sim 1 + \frac{\sqrt{K_{\text{diff},\omega} t}}{R_{NP}}$$

and

$$\alpha^{\text{dynamic}}_{\text{temp gradient}} \sim \frac{\Delta T_{\max}(t)}{I_{\text{flux}}} \frac{1}{\sqrt{K_{\text{diff},\omega} t}},$$

where $K_{\text{diff},\omega}$ is the thermal diffusivity. These equations are valid for $t > \Delta t_{\text{pulse}}$, where Δt_{pulse} is the length of the pulse. In addition, $\alpha^{\text{dynamic}}_{\text{localization length}} < \alpha^{\text{static}}_{\text{localization length}}$ and $\alpha^{\text{dynamic}}_{\text{tempgradient}} > \alpha^{\text{static}}_{\text{tempgradient}}$ due to the finite time that it takes the NP to transfer the heat. Therefore, for short times, the pulse excitation results in a stronger localized temperature increase than the CW regime. This is because of the temperature increase drop described in Eq. (2.11). In contrast, for long times the degree of localization of the temperature is greatly reduced due to the spreading of the hot spot.

References

1. Noguez C (2007) Surface plasmons on metal nanoparticles: the influence of shape and physical environment. J Phys Chem C 111(10):3806–3819
2. Bohren CFH (1983) Absorption and scattering of light by small particles. Wiley, New York
3. Choy TC (1999) Effective medium theory: principles and applications. Oxford University Press, New York
4. Govorov AO, Zhang H, Demir HV, Gun'ko YK (2014) Photogeneration of hot plasmonic electrons with metal nanocrystals: quantum description and potential applications. *Nano Today* 9(1):85–101
5. Garcia MA (2012) Surface plasmons in metallic nanoparticles: fundamentals and applications. J Phys D Appl Phys 45(38):389501
6. Kelly KL, Coronado E, Zhao LL, Schatz GC (2003) The optical properties of metal nanoparticles: the influence of size, shape, and dielectric environment. J Phys Chem B 107 (3):668–677
7. Khosravi Khorashad L, Besteiro LV, Wang Z, Valentine J, Govorov AO (2016) Localization of excess temperature using plasmonic hot spots in metal nanostructures: combining nano-optical antennas with the fano effect. J Phys Chem C 120(24):13215–13226
8. Johnson PB, Christy RW (1972) Optical-constants of noble-metals. Phys Rev B 6(12):4370–4379
9. Palik ED (1985) Handbook of optical constants of solids. Academic Press, New York
10. Carslaw HS, Jaeger JC (1993) Conduction of heat in solids. Oxford University Press, London
11. Pitsillides CM, Joe EK, Wei XB, Anderson RR, Lin CP (2003) Selective cell targeting with light-absorbing microparticles and nanoparticles. Biophys J 84(6):4023–4032
12. Govorov AO, Richardson HH (2007) Generating heat with metal nanoparticles. Nano Today 2(1):30–38
13. Sau TK, Rogach AL (2012) Complex-shaped metal nanoparticles. Verlag & Co. KGaA, Weinheim, Germany
14. Govorov AO, Zhang W, Skeini T, Richardson H, Lee J, Kotov NA (2006) Gold nanoparticle ensembles as heaters and actuators: melting and collective plasmon resonances. Nanoscale Res Lett 1(1):84–90
15. Hrelescu C, Stehr J, Ringler M, Sperling RA, Parak WJ, Klar TA, Feldmann J (2010) DNA melting in gold nanostove clusters. J Phys Chem C 114(16):7401–7411
16. Richardson HH, Hickman ZN, Govorov AO, Thomas AC, Zhang W, Kordesch ME (2006) Thermooptical properties of gold nanoparticles embedded in ice: characterization of heat generation and melting. Nano Lett 6(4):783–788
17. Zeng N, Murphy AB (2009) Heat generation by optically and thermally interacting aggregates of gold nanoparticles under illumination. Nanotechnology 20(37):375702
18. Roller E-M, Besteiro LV, Pupp C, Khorashad LK, Govorov AO, Liedl T (2017) Hotspot-mediated non-dissipative and ultrafast plasmon passage. Nat Phys 13:761

Chapter 3
Nanoscale Temperature Measurement Under Optical Illumination Using AlGaN:Er^{3+} Photoluminescence Nanothermometry

Susil Baral, Ali Rafiei Miandashti and Hugh H. Richardson

3.1 Introduction

Nanoscale heat generation and dissipation impacts many fields of current research, including applications in medical therapies and the semiconductor industry, where device dimensions continue to be reduced as number densities increase. As device sizes approach the nanoscale, they become close in scale to the phonon mean-free path, where certain pathways for heat dissipation are less efficient. The classical heat diffusion law breaks down and the heat flow becomes ballistic [1–3]. Also in this regime, interfacial properties begin to dominate and limit the heat transfer away from the nanostructure [4]. Characterizing these interfacial properties and their impact upon heat dissipation is essential for an understanding of nanoscale heat transport. Novel optical thermal sensor can be used to measure the local temperature of a thin film. One of the most recent optical thermal sensors are luminescent films. The optical thermal sensor is made of a thin film of $Al_{0.94}Ga_{0.06}N$ incorporated with Er^{3+}. The sensor is used to determine the temperature of the film around either a single 40 nm gold nanoparticle (NP) or a single lithographically prepared nanodot during laser excitation.

3.2 AlGaN:Er^{3+} Photoluminescence Nanothermometry

A thin film of $Al_{0.94}Ga_{0.06}N$ with embedded Er^{3+} ions is used as a thermal sensor for nanoscale temperature measurement and thermal imaging [5]. This thermal sensor film use relative photoluminescence intensities of the $^2H_{11/2} \rightarrow {}^4I_{15/2}$ and the $^4S_{3/2} \rightarrow {}^4I_{15/2}$ energy transitions of the Er^{3+} ions. These intensities have been shown to be temperature dependent [6, 7] and are related by a Boltzmann factor (exp($-\Delta E$/kT) where ΔE is the energy difference between the two levels, k is the Boltzmann constant, and T is the absolute temperature.

A. R. Miandashti et al., *Photo-Thermal Spectroscopy with Plasmonic and Rare-Earth Doped (Nano)Materials*, Nanoscience and Nanotechnology,
https://doi.org/10.1007/978-981-13-3591-4_3

The energy level diagram of Er^{3+} embedded on AlGaN matrix and the corresponding photoluminescence from Er^{3+} excited states to the ground state is presented in Fig. 3.1a. Erbium ion has several energy levels in visible and near IR region of an electromagnetic radiation. When the Er^{3+} doped thermal sensor film is excited at 532 nm, the energy is first absorbed by the defect energy states of a semiconductor AlGaN. The energy is then transferred (vibrationally relaxed) into Er^{3+} energy levels, which results in the excitation of two closely spaced Er^{3+} levels, namely $^2H_{11/2}$ and $^4S_{3/2}$. The photoluminescence from these two energy states to the ground state ($^4I_{15/2}$) takes place at approximately 540 and 560 nm respectively.

As shown in the energy level diagram (Fig. 3.1), $^2H_{11/2}$ and $^4S_{3/2}$ excited energy levels are close to each other with relatively small energy gap (800 $cm^{-1)}$ between them and hence are in quasi-thermal equilibrium. Room temperature excitation permits the low energy $^4S_{3/2}$ level to populate with respect to the $^2H_{11/2}$ level under proper excitation; and as the temperature increases, the high energy $^2H_{11/2}$ level is increasingly populated from the $^4S_{3/2}$ level. The relative population and hence the photoluminescence intensity from these "thermally coupled" energy levels is, therefore, temperature dependent and follows a Boltzmann type population distribution between the energy levels. As each of these energy levels is associated with their own temperature dependence, temperature can be inferred by measuring the intensity of the photoluminescence from a particular level. But the drawback of this technique is that any fluctuations on intensity could be misinterpreted as temperature change, leading to higher error on temperature measurement. Fluorescent Intensity Ratio (FIR) technique helps to avoid such misinterpretation and error by measuring the intensity of two closely spaced energy levels and taking the ratio to measure the temperature. Because the emission intensities are proportional to the

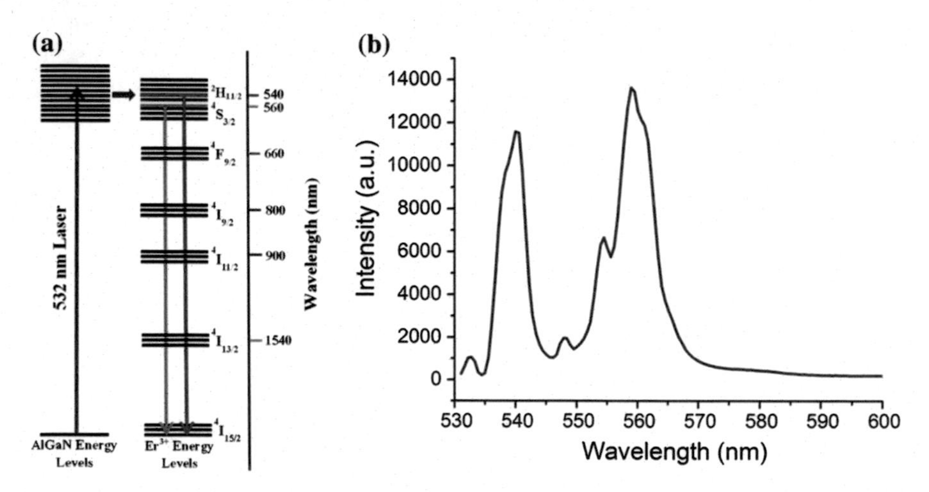

Fig. 3.1 Energy diagram of Er^{3+} embedded on AlGaN matrix and the corresponding temperature dependent photoluminescence from Er^{3+} excited states to the ground state. Reprinted with permission from reference [11]. (Energy levels not in scale)

population of each energy level, the FIR from two thermally couple energy levels is given as: [5, 8, 9]

$$FIR = \frac{N_i}{N_j} = \frac{I_i}{I_j} = \frac{g_i\sigma_i\omega_i}{g_j\sigma_j\omega_j}\exp\left[\frac{-\Delta E}{kT}\right] = B\exp\left[\frac{-\Delta E}{kT}\right]$$
$$where, B = \frac{g_i\sigma_i\omega_i}{g_j\sigma_j\omega_j} \quad (3.1)$$

In Eq. (3.1); N, I, g, σ, ω represents the number of ions, the fluorescence intensity, the degeneracy, the emission cross section, and the angular frequency of fluorescence transitions from the upper (i) and lower (j) thermally couple energy levels respectively. ΔE is the energy difference between the two thermally coupled levels,k is the Boltzmann constant and T is the temperature in absolute scale [5, 9].

3.3 Experimental Details of AlGaN:Er^{3+} Photoluminescence Nanothermometry

In a typical experiment, a sample consists of nanoheaters (gold nanoparticles/nanorods/nanostructures) on top of a thermal sensor film of $Al_{0.94}Ga_{0.06}N:Er^{3+}$ on Si or Sapphire substrate. A sample can be prepared by drop casting/spin-coating dilute colloidal solution of gold nanoheaters (nanoparticles/nanorods) on top of a thermal sensor film. More specific nanostructures can be fabricated on top of a thermal sensor film by various lithographic techniques such as hole-mask colloidal lithography/soft lithography/nanosphere lithography (NSL), electron beam lithography.

Optical collection of photoluminescence from the AlGaN:Er^{3+} thermal sensor film and hence the temperature measurement is performed with Scanning Near Field Optical Microscope (WITec α-SNOM300s). The microscope (pictured below) has imaging and spectral measurement capabilities under Near-field, Far-field, Dark-field, Confocal and Raman/Photoluminescence mode. In addition, the microscope can also perform AFM imaging measurements. Representative picture of a microscope and a typical schematic optical setup for measurement under Far-field Raman/Photoluminescence mode is shown below.

Typical optical measurement of temperature is performed on a Raman/Photoluminescence mode of a SNOM under far-field imaging mode. A sample is first placed on the sample stage and is focused with white light brought in from the side of the microscope through an objective lens (typically 50X, 0.8 Nikon Air Objective in air; and 60X, 1.0 Canon Water Immersion Objective in water). An adjustable 532 nm CW Nd:YAG laser is also brought in from the side of the microscope which hits a corner cube and is focused onto the sample, as illustrated in the block diagram shown in Fig. 3.2. The laser excites the thermal sensor film and the Er^{3+} emission from the film is collected in reflection mode with the same

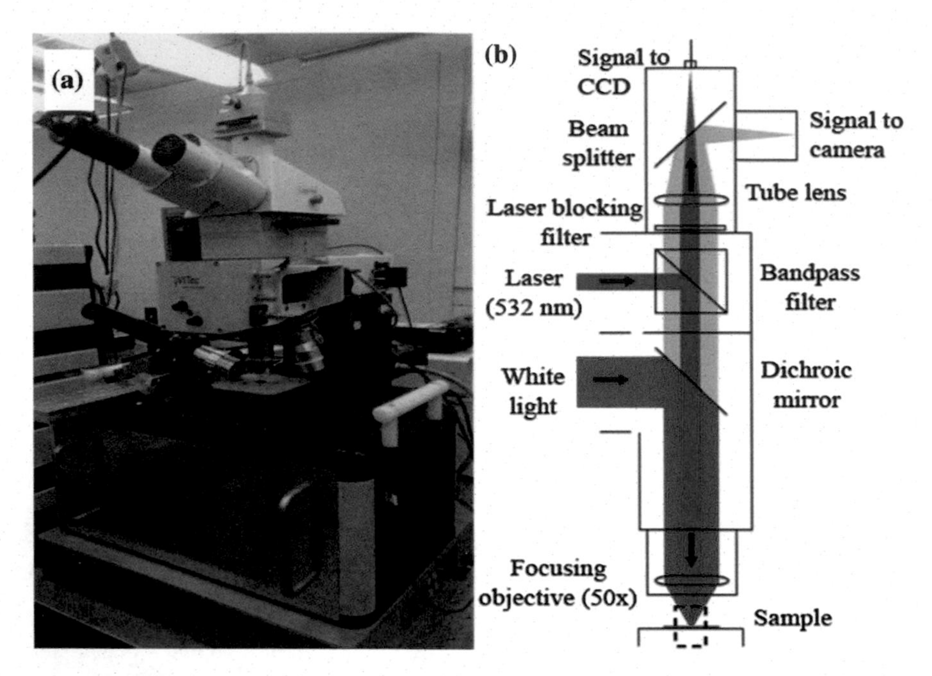

Fig. 3.2 **a** Picture of a scanning near-field optical microscope (SNOM) and **b** schematic optical setup of the microscope used for optical measurements. Reprinted with permission from reference [11].

objective and sent to CCD spectrograph via a collection fiber (typically 25 μm). Before reaching the collection fiber, the reflected emission from the film is passed through the band pass filter (Long pass, 532 nm) which cuts off the excitation laser and protects the CCD. A typical Er^{3+} photoluminescence spectrum from AlGaN: Er^{3+} collected with SNOM microscope is presented in Fig. 3.1b.

The microscope (SNOM) has a piezo stage which offers scanning and hence performing imaging measurements. This enables to generate an image and the corresponding photoluminescence spectrum at different points in space for a predetermined integration time, pixel size, and image size. The imaging measurement is performed by scanning the substrate (thermal sensor film of AlGaN:Er^{3+} on Silicon/Sapphire) with gold nanoparticles/nanostructures under the laser and collecting photoluminescence from Er^{3+} continuously throughout the scan. While imaging, only the region of the substrate under laser illumination is excited/heated and only the photoluminescence information from the same region is collected for the temperature measurement. When the laser is scanning over the gold nanoparticle/nanostructure, the particle will generate heat and transfer the heat into the thermal sensor film which is also under the excitation laser illumination. The temperature increase (change) will be collected and stored in the corresponding spectra collected form that region. On the other hand, when the laser is scanning aside the nanostructure (the region with film only and no nanoparticle/ nanostructure) the nanostructure is no longer hot and the temperature is basically the room temperature, which is again collected and stored in the corresponding spectra collected form that region. At the end of a typical imaging measurement, a photoluminescence image and the corresponding Er^{3+} photoluminescence spectrum is obtained. The temperature (thermal) images are then constructed by spatially analyzing the relative peak intensities of the Er^{3+} emission from the film collected throughout the scan. The figure below shows a typical imaging measurement performed with SNOM and Er^{3+} photoluminescence spectra at 'hot' and 'cold' regions on the substrate corresponding optically heated gold nanoparticle and background respectively (Fig. 3.3).

After temperature image (thermal image) is constructed, a cross section is drawn across the hot spot to generate the corresponding thermal profile of the optically heated nanoheater. The measured temperature of the nanoheater is the difference between the maximum temperature of the nanoheaters and the background temperature as illustrated in the Fig. 3.4a, b.

The optical temperature measurement using AlGaN:Er^{3+} film is resolution limited and needs to be convoluted with the collection volume of the microscope and the true thermal image in the substrate. The thermal transfer parameter was determined in order to convert the measured temperature to the local temperature of the nanostructure under excitation. The local temperature change expected for nanostructure excitation under certain laser intensity was calculated with the help of Mie Theory and is given as shown in Eq. (3.2), [5, 9, 10]

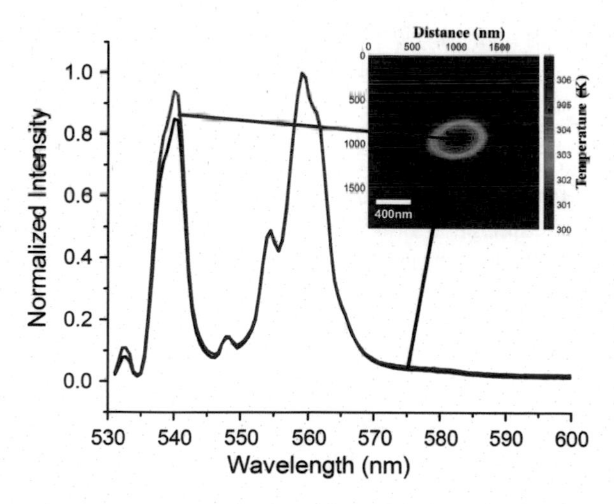

Fig. 3.3 Er^{3+} photoluminescence spectrum from different regions of AlGaN:Er^{3+} on Silicon substrate. The black spectrum represents the background (cold region) where the intensity of low energy band ^4S band is higher than that for the high energy ^2H band. The red spectrum represents the hot region with optically heated gold nanostructure where the intensity of ^2H band is increased with respect to that of ^4S band because of the temperature dependent population and hence the emission of thermally coupled H and S energy levels. Reprinted with permission from reference [11].

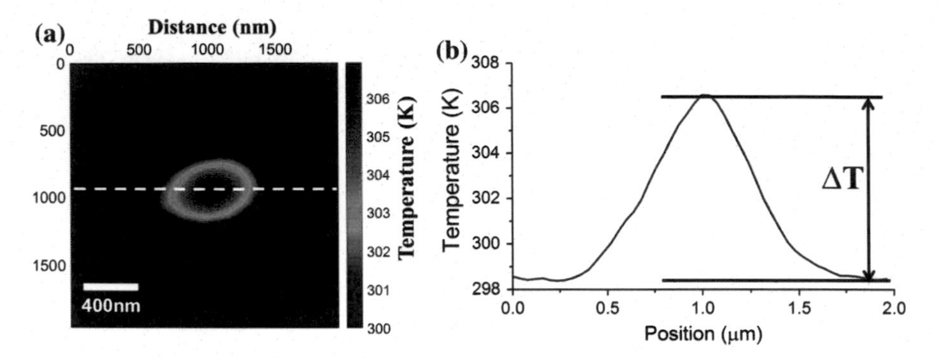

Fig. 3.4 **a** Thermal image of a rod-shape nanoheater on AlGaN:Er^{3+} on Silicon. **b** The corresponding thermal profile for a cross section drawn through the hot spot of a nanoheater. Reprinted with permission from reference [11].

$$\Delta T = \frac{C_{abs} I_0}{4\pi k_{eff} R_{NP}} \tag{3.2}$$

where C_{abs} is the absorption cross section of the nanostructure, I_0 is the excitation laser intensity, k_{eff} is the thermal conductivity of the medium and R_{NP} is the radius of the nanoparticle. A scaling factor (thermal transfer parameter) was then determined so that the measured temperature when multiplied by the factor gives temperature change (ΔT) as expected from Eq. (3.2). The value of the thermal transfer coefficient was determined to be 12.8 for nanostructure excitation in air (using 50× objective lens) and 10.2 for that on water (using 60× water immersion lens) because of the different numerical aperture associated with different collection lens. These parameters were determined by careful thermal calibration of the AlGaN:Er^{3+} film and the detail descriptions can be found in the literature [5].

In addition to the imaging measurements, the instrument (SNOM) also allows making dynamic measurements of temperature on a selected nanoheater for desired period of time in the form of time spectrum. For a typical dynamic measurement, a survey spectral image is first taken to locate the nanoheater. Once the position of the nanoheater is located, a background time spectrum is first taken with laser excitation over a period of time while focused onto the background (the region with no nanostructure). A time spectrum is then taken while focused onto the nanostructure for a desired period of time. The temperature spectrum is constructed from photoluminescence spectrum collected for both the background and the nanostructure excitation. The temperature difference spectrum is then generated by taking the difference between the time temperature spectrum for nanoheater and the background. An example of a typical temperature-time measurement is shown in a Fig. 3.5.

Dynamic measurements of temperature can also be performed with increasing laser intensity with time while focused onto the nanostructure for a period of time as in Fig. 3.6.

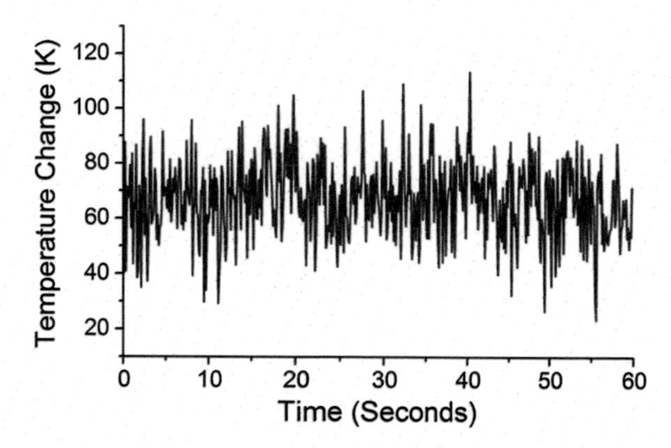

Fig. 3.5 Typical time spectrum showing temperature change on nanoheater as a function of a time for a laser power of $\sim 10^8$ W/m^2 for a gold nanodisk with radius of $\sim$300 nm and height of $\sim$100 nm and sampling rate of $\sim$20 ms. Reprinted with permission from reference [11].

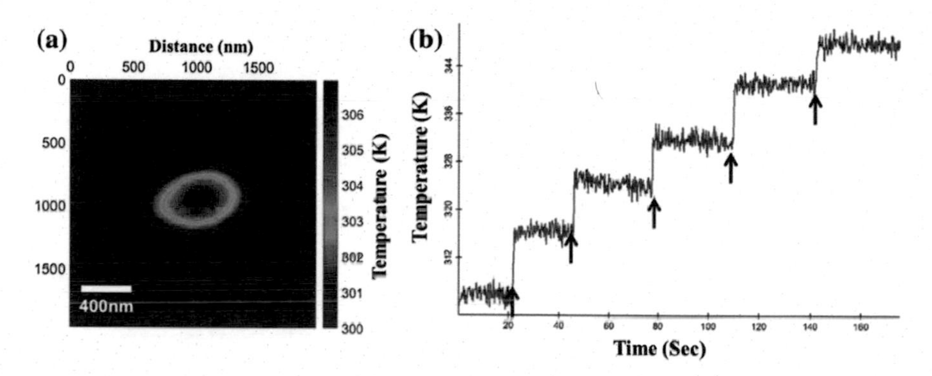

Fig. 3.6 Temperature-time spectrum showing temperature change on a rod-shape nanoheater as a function of a time with increasing laser intensity from 10^8 to 10^9 W/m^2. Black arrows designate the regions where laser intensity was increased

References

1. Chen G (2001) Ballistic-diffusive heat-conduction equations. Phys Rev Lett 86(11):2297–2300
2. Chen G (1996) Nonlocal and nonequilibrium heat conduction in the vicinity of nanoparticles. J Heat Transf Trans Asme 118(3):539–545
3. Siemens ME, Yang R, Nelson KA, Anderson EH, Murnane MM, Kapteyn HC (2010) Quasi-ballistic thermal transport from nanoscale interfaces observed using ultrafast coherent soft x-ray beams. Nat Mater 9:26–30
4. Cahill DG, Ford WK, Goodson KE, Mahan GD, Majumdar A, Maris HJ, Merlin R, Phillpot SR (2003) Nanoscale thermal transport. J Appl Phys 93(2):793–818
5. Carlson MT, Khan A, Richardson HH (2011) Local temperature determination of optically excited nanoparticles and nanodots. Nano Lett 11(3):1061–1069

6. Gurumurugan K, Chen H, Harp GR, Jadwisienczak WM, Lozykowski HJ (1999) Visible cathodoluminescence of Er-doped amorphous AlN thin films. Appl Phys Lett 74(20):3008–3010
7. Garter MJ, Steckl AJ (2002) Temperature behavior of visible and infrared electroluminescent devices fabricated on erbium-doped GaN. IEEE Trans Elect Dev 49(1):48–54
8. Paez G, Strojnik M (2003) Erbium-doped optical fiber fluorescence temperature sensor with enhanced sensitivity, a high signal-to-noise ratio, and a power ratio in the 520–530- and 550–560-nm bands. Appl Opt 42(16):3251–3258
9. Wade SA, Collins SF, Baxter GW (2003) Fluorescence intensity ratio technique for optical fiber point temperature sensing. J Appl Phys 94(8):4743–4756
10. Govorov AO, Zhang W, Skeini T, Richardson H, Lee J, Kotov NA (2006) Gold nanoparticle ensembles as heaters and actuators: melting and collective plasmon resonances. Nanoscale Res Lett 1(1):84–90
11. Baral S (2017) Fundamental Studies of Photothermal Properties of a Nanosystem and the Surrounding Medium Using Er^{3+} Photoluminescence Nanothermometry. (Electronic Thesis or Dissertation). Retrieved from https://etd.ohiolink.edu/

Chapter 4
Comparison of Nucleation Behavior of Surrounding Water Under Optical Excitation of Single Gold Nanostructure and Colloidal Solution

Susil Baral, Ali Rafiei Miandashti and Hugh H. Richardson

4.1 Introduction

Colloidal solution of gold nanoparticles optically excited form steam bubbles at the boiling point. In colloidal solution, the ambient liquid temperature is raised to the boiling point by collective heating and boiling occurs within the solution and not necessarily at the surface of the nanoparticle. This mechanism is consistent with classical nucleation theory where boiling in a bulk liquid starts at a nucleation center (a small air bubble or another object). Nucleation centers diffuse in a liquid and sometimes enter the superheated area. Therefore, the probability and frequency of appearance of vapor bubbles strongly depend on the volume of the superheated water. In the case of the solution experiments, this superheated-water volume is orders of magnitude larger than that in the single nano-wrench experiment. This extreme difference in heated volumes leads to the behavior where vapor bubble creation in the solution case is a frequent event, whereas, in the single nanostructure case, bubbles happen to be very rare events and happen at much higher temperatures.

4.2 Temperature Changes and Phase Transformation with Gold Nano-wrenches

Single gold nano-wrenches (fabricated with e-beam lithography on top of a thermal senor AlGaN:Er^{3+} film on the Silicon substrate) are optically excited using 532 nm CW laser and the corresponding temperature change on the nanostructure is measured from Er^{3+} photoluminescence intensity (as described in the previous chapter) and plotted as a function of an excitation laser intensity (shown in Fig. 4.1).

A. R. Miandashti et al., *Photo-Thermal Spectroscopy with Plasmonic and Rare-Earth Doped (Nano)Materials*, Nanoscience and Nanotechnology,
https://doi.org/10.1007/978-981-13-3591-4_4

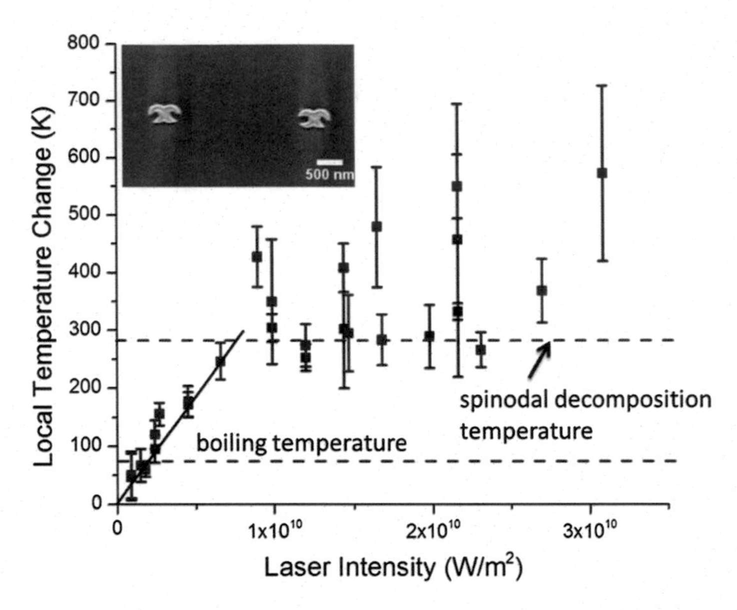

Fig. 4.1 A plot of a local temperature change in K as a function of excitation laser intensity in W/m^2. Inset shows the SEM image of a typical gold nano-wrenches. The nano-wrenches were fabricated using e-beam lithography in arrays with spacing between nano-wrenches of 3000 nm. The beam width of the excitation laser (FWHM) is $\sim$800 nm insuring that only a single gold nano-wrench is excited. Different colored data points in the figure represent data collected on different days. Reprinted with permission from *ACS Nano* 2014, *8*(2), 1439–1448. Copyright 2014 American Chemical Society

The data shown in Fig. 4.1 can be divided into two different regions where the properties are quite different. At low laser intensities, a linear region is observed where the local temperature varies linearly with laser intensity. The linear temperature region goes beyond the boiling point of water (373 K) and extends to the spinodal decomposition temperature of water at 580 K. At the spinodal decomposition temperature of water, the characteristics of temperature with laser intensity changes dramatically. In the second region the temperature does not vary linear with laser intensity, but multiple temperatures are observed at each laser intensity.

4.3 Dynamic Temperature Changes and Phase Transformation with Gold Nano-wrenches

The temperature variation shown in Fig. 4.2 is probed by taking temperature time traces at constant laser intensity. Figure 4.2a shows temperature time traces at different laser intensities. At low laser intensity, the average local temperature is below the boiling point of water. In temperature trace A, a single nano-wrench was heated with a laser intensity of 1.5×10^9 W/m^2 for a time period of 60 s.

The photoluminescence spectrum at each time interval is converted into temperature. Background temperature traces are collected for each laser intensity and subtracted from the sampled temperature trace to give the difference temperature trace shown in Fig. 4.2a. The number of counts in the photoluminescence spectrum scales with laser intensity so a lower temperature variation is expected at higher laser intensities. Temperature traces B (laser intensity 2.4×10^9 W/m^2) and C (laser intensity 4.5×10^9 W/m^2) are at an average temperature above the boiling point of water but below the spinodal decomposition temperature. The temperature variation in the trace is diminishing with higher laser intensity as expected. This behavior changes in temperature trace D (laser intensity 1.2×10^{10} W/m^2) taken at the spinodal decomposition temperature of 580 ± 20 K. The variation in temperature has both high and low frequency components. Above the spinodal decomposition temperature the temperature variation increased with temperature (traces E and F, laser intensity 1.44×10^{10} and 2.16×10^{10} W/m^2 respectively). The variation in temperature as a function of temperature can be characterized by taking the standard deviation of the temperature trace from the average temperature and multiply by the laser intensity (shown in Fig. 4.2b). If the system is behaved as expected, then such a plot should show no variation with temperature. Figure 4.2b shows the standard deviation multiplied by the laser intensity as a function of the local temperature change. A flat line is observed at low temperature that slope increases to the spinodal decomposition temperature where a dramatic break in the trend is observed. At temperatures at and above the spinodal decomposition

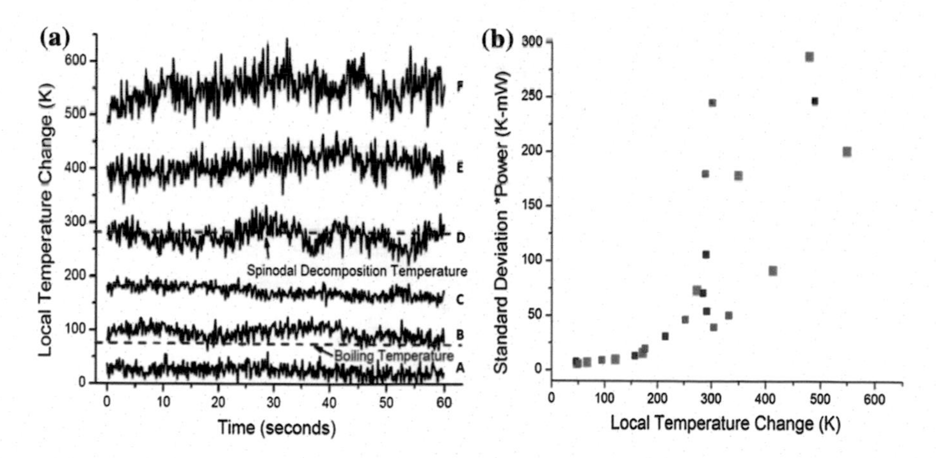

Fig. 4.2 **a** Temperature-time spectra of a single nano-wrench at different laser intensities. The nano-wrench was heated with different laser intensities (A: 1.50×10^9, B: 2.40×10^9, C: 4.50×10^9, D: 1.20×10^{10}, E: 1.44×10^{10}, F: 2.16×10^{10} W/m^2) for a time period of 60 s. The red dashed line designates the boiling and spinodal decomposition temperature of water. **b** Plot of standard deviation multiplied by the laser intensity as a function of the local temperature change. The standard deviation is determined from temperature-time spectra shown in (**a**). Reprinted with permission from *ACS Nano* 2014, *8*(2), 1439–1448. Copyright 2014 American Chemical Society

temperature ($\Delta T = 280 \pm 20$ K), wild fluctuations in temperature are observed that increase the standard deviation significantly. In order to probe the temperature structure properties of the water near the nano-wrench, both the lossless scattering of the excitation laser and photoluminescence from the thermal sensor film were collected simultaneously.

The overlaid temperature and scattering time traces are shown in Fig. 4.3 for temperatures between the boiling and spinodal decomposition points and for temperatures much above the spinodal decomposition temperature. The data, shown in red, is scattering collected back through the microscope objective at the excitation wavelength. The data, shown in black, is the temperature trace converted from the photoluminescence spectrum and scaled to yield the local temperature. The data shown in A profile of Fig. 4.3 is taken at a laser intensity of 4.4×10^9 W/m^2. The temperature of the nano-wrenches for this laser intensity is near the boiling point of the water and both the temperature and scattering traces have similar properties. In contrast to this behavior, data shown in B profile indicate a temperature much above the spinodal decomposition temperature (laser intensity 3×10^{10} W/m^2), the temperature and scattering profiles move in opposite directions with an increase in temperature corresponding to a decrease in laser scattering.

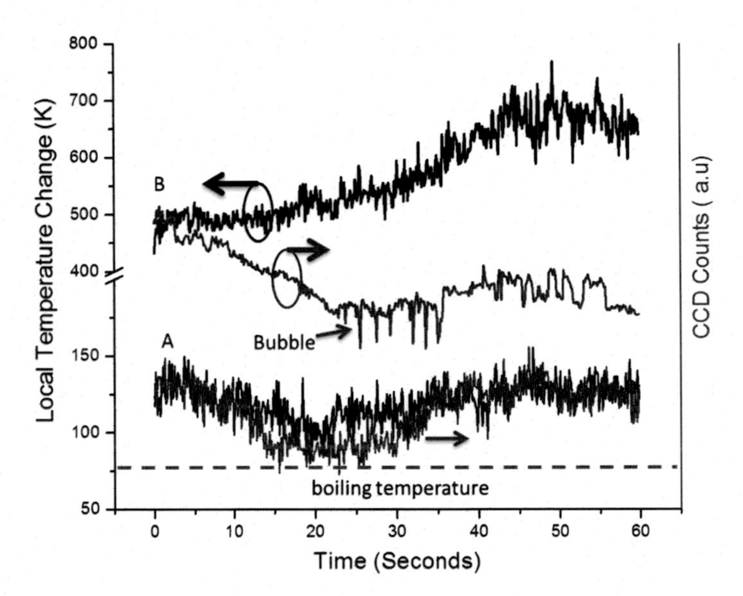

Fig. 4.3 Temperature-time spectra (shown in black) overlaid on laser scattering spectra (shown in red) at laser intensity of 2.4×10^9 W/m^2 and 3.10×10^{10} W/m^2. Evidence of bubble formation is revealed as dips in the laser scattering spectra. Bubbles formed at temperatures between the boiling and spinodal decomposition temperature are rare but long lived while bubble formation for temperatures above the spinodal decomposition temperature are shorter lived but happen frequently. Reprinted with permission from *ACS Nano* 2014, *8*(2), 1439–1448. Copyright 2014 American Chemical Society

4.4 Temperature Measurements of Optically Excited Colloidal Gold Nanoparticles

A gold nanoparticle colloidal solution, concentration of 1×10^{11} particles/cm^3 and 13 nm diameter, is excited at 532 nm with 1.0 W laser power. The laser spot size is 0.3 mm yielding a laser intensity of 1.8×10^7 W/m^2. Figure 4.4 shows three consecutive frames, each 0.04 s apart, taken from a movie of water boiling when excited with laser light. The green line in the figure is the 532 nm laser. Panel a is just before boiling while panels b and c show spontaneous boiling of the solution with a large bubble forming within the capillary tube.

4.5 Temperature Measurements Probing Convection of the Liquid During Laser Excitation of a Colloidal Nanoparticle Solution

Figure 4.5 shows the temperature inside the capillary tube at distances away from the laser excitation area for a 40 nm diameter gold colloidal solution of concentration 9×10^{10} particles/cm^3. The laser power is changed from 0 to 500 mW and the corresponding temperature is measured using the thermocouple. The temperature within the laser excitation (0 mm distance) is impossible to probe with the thermocouple and is extrapolated from the exponential fit of the thermal data away from the center of the laser excitation area.

The temperature at the center of the laser excitation area has been estimated for laser excitation of a gold colloidal droplet [1] (Eq. 4.1) or for a spherical region with randomly distributed nanoparticle heat sources [2] (Eq. 4.2). Equation 4.2 gives the temperature increase of a heater with a characteristic size R_{heat} and it is

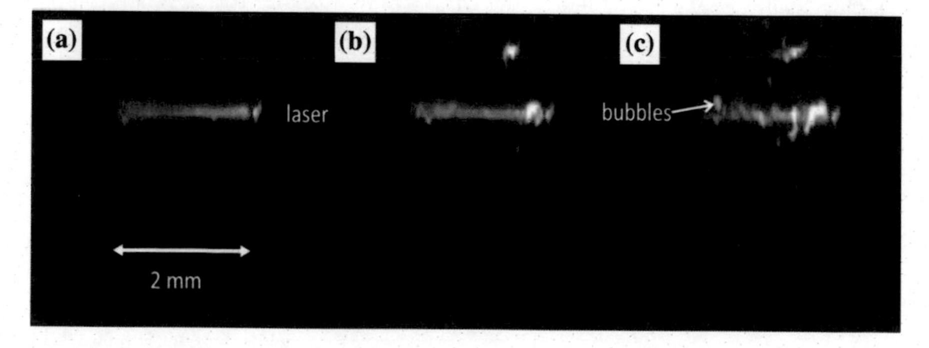

Fig. 4.4 Three successive frames each 0.04 s apart from a movie showing water boiling after laser excitation of a colloidal solution containing 13 nm diameter gold nanoparticles. The solution is excited at 532 nm with a laser intensity of 1.8×107 W/m^2. Reprinted with permission from *ACS Nano* 2014, *8*(2), 1439–1448. Copyright 2014 American Chemical Society

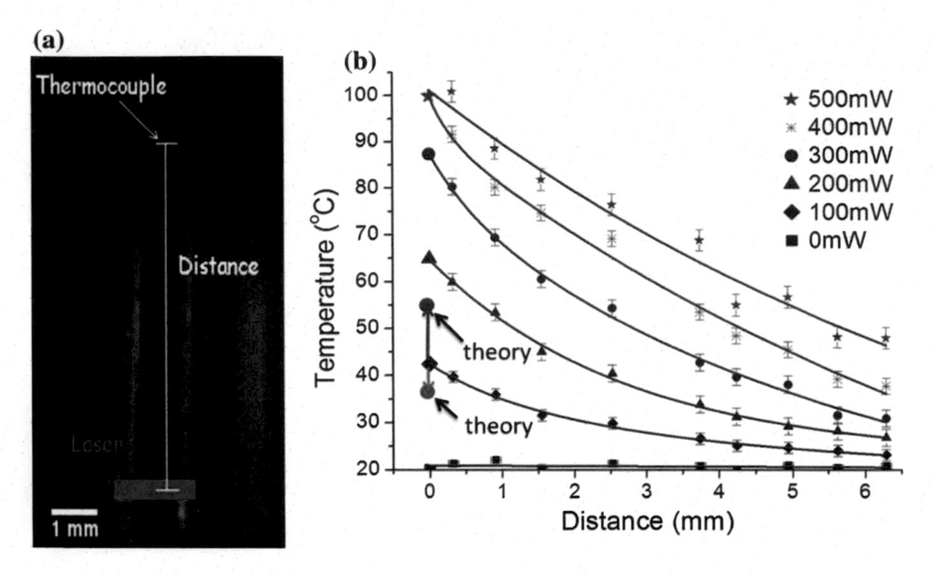

Fig. 4.5 Temperature profiles of the liquid within the capillary tube (2 mm diameter) at distances away from the laser excitation area when a gold nanoparticle colloidal solution, concentration of 9×10^{10} particles/cm^3 and 40 nm diameter, is excited at 532 nm. The laser intensity is changed from 0 to 9×10^6 W/m^2. Two theoretical points calculated by Eqs. (4.1) and (4.2) are also included for the 100 mW excitation. Reprinted with permission from *ACS Nano* 2014, *8*(2), 1439–1448. Copyright 2014 American Chemical Society

valid relatively far from the cylinder surface. In Eqs. (4.1) and (4.2), C_{abs} is the absorption cross section of a 40 nm gold nanoparticle immersed in water (2.8×10^{-15} m^2), I (W/m^2) is the intensity of the excitation laser [Power (W)] 1.8×10^7 m^{-2}), R_{NP} is the gold nanoparticle radius (20 nm), η_{NP} is the nanoparticle concentration (9×10^{10} particles/cm^3), l_{opt} is the optical path length (1.649 mm), R_{beam} is the radius of the optical beam (0.136 mm) [1]. Theory (Eq. 4.1) generally underestimates the temperature for the 100 mW excitation, shown as the large blue point in Fig. 4.5, while theory from a spherical heated region (Eq. 4.2) overestimates the temperature (shown as the large red point). The thermal profile away from the cylinder of laser excitation has been previously calculated [1, 3]. The damping of the temperature away from the excitation region has an exponential decay but at a larger damping rate than the experimental thermal profiles shown in Fig. 4.5b. Theory overestimates the damping of the temperature profile away from the excitation region because thermal diffusion due to convection is neglected.

$$\Delta T \approx \frac{C_{abs} I \eta_{NP} R_{beam}^2 \log_e(l_{opt}/R_{beam})}{2k_w} \tag{4.1}$$

$$\Delta T_{global} = \frac{C_{abs} I \eta_{NP} R_{heat}^2}{2k_w} \tag{4.2}$$

Convection within the liquid is probed by measuring the temperature distribution within the optically stimulated liquid as the laser is moved from the bottom to top of a liquid sample cell. The temperature profiles within the liquid at different laser positions are shown in Fig. 4.6c. The color-coded arrows show the position of the laser excitation. When the sample is excited in the middle of the sample cell (laser position 3, red data), the temperature in the liquid 2 mm above the laser spot is higher than the temperature at the laser spot. Conversely, the temperature in the liquid 2 mm below the laser spot is cooler than the temperature at the laser spot. This trend is observed for all the laser positions except for position 5 where the sample is excited near the bottom of the sample cell. This effect is consistent with fluid motion through convection consistent with observation from micrometer sized silica beads in laser excited gold colloidal droplets [4].

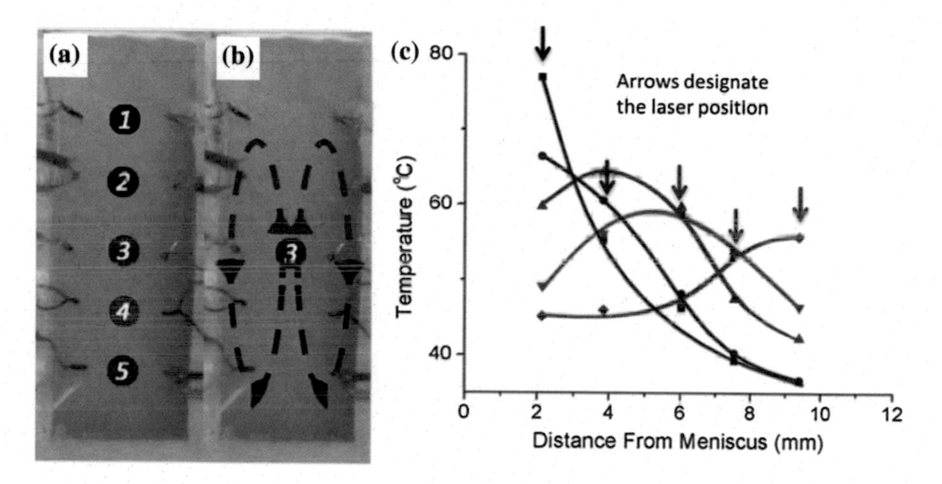

Fig. 4.6 **a** Picture of the sample cell showing the position of laser excitation and the thermocouples used to measure the temperature within the sample cell. **b** Diagram showing the expected rise in hot liquid when excited at laser position 3. **c** The temperature profiles within the liquid at different laser positions. The color-coded arrows show the position of the laser excitation. When the sample (40 nm gold nanoparticle solution) is excited (532 nm, 1.25 W) in the middle of the sample cell (laser position 3, red data), the temperature in the liquid 2 mm above the laser spot is higher than the temperature at the laser spot indicating movement of the hot liquid. Reprinted with permission from *ACS Nano* 2014, *8*(2), 1439–1448. Copyright 2014 American Chemical Society

References

1. Richardson HH, Carlson MT, Tandler PJ, Hernandez P, Govorov AO (2009) Experimental and theoretical studies of light-to-heat conversion and collective heating effects in metal nanoparticle solutions. Nano Lett 9(3):1139–1146
2. Keblinski P, Cahill DG, Bodapati A, Sullivan CR, Taton TA (2006) Limits of localized heating by electromagnetically excited nanoparticles. J Appl Phys 100:(5)
3. Govorov AO, Zhang W, Skeini T, Richardson H, Lee J, Kotov NA (2006) Gold nanoparticle ensembles as heaters and actuators: melting and collective plasmon resonances. Nanoscale Res Lett 1(1):84–90
4. Carlson MT, Barton TS, Tandler PJ, Richardson HH, Govorov AO (2009) Thermal effects of colloidal suspensions of Au nanoparticles. Mater Res Soc Symp Proc 1172:T05-08

Chapter 5
Effect of Ions and Ionic Strength on Surface Plasmon Extinction Properties of Single Plasmonic Nanostructures

Susil Baral, Ali Rafiei Miandashti and Hugh H. Richardson

5.1 Introduction

In previous reports, it is observed that temperature drops in a thin film used as a temperature sensor when single gold nanowires are optically excited and the liquid solution at the interface is changed from pure water to aqueous solutions of dilute glucose or NaCl [1]. It was reported that the temperature drop is due to an increase in the heat dissipation into the solution that is controlled by changes in the interface. This assignment stands on the assumption that there is no change in the heat generation of the optically excited gold nanowire when a dilute concentration of ions or glucose is added to water because the change in the solution dielectric constant is negligible and hence no change in the plasmon absorption is expected. Two recent papers present results that appear to question this assignment. The first paper observed attenuation of plasmon absorption from thin gold films at high ionic concentrations of ionic liquids (1–8 molar concentrations of KCl, NaCl, Na_2SO_4, and $MgCl_2$) [2]. Also, another paper presents results that a shift in the plasmon absorption band for gold nanodisks is observed at a threshold concentration of 10^{-2} mM glucoside (for hydrophobic Au) and 1 mM glucoside (for hydrophilic Au). An approximate 6 nm plasmon shift was observed for hydrophobic Au nanodisks in a 100 mM solution of glucoside. This observed shift is attributed to an interfacial effect that magnifies the plasmon shift over what is expected based upon changes in the surrounding dielectric constant [3]. The authors also found that adding a dilute solution of glucoside does not change the rate of heat dissipation in the solution compared to pure water.

A. R. Miandashti et al., *Photo-Thermal Spectroscopy with Plasmonic and Rare-Earth Doped (Nano)Materials*, Nanoscience and Nanotechnology,
https://doi.org/10.1007/978-981-13-3591-4_5

5.2 Measurement of Nanoscale Temperature Change on Optically Excited Gold Nanowires Using AlGaN: Er^{3+} Nanothermometry

The local temperature change under 532 nm CW excitation of a single gold nanowire (fabricated using e-beam lithography) in air and aqueous solutions at concentrations from 0 to 0.2 mol/L for (a) NaCl, (b) Na_2SO_4 and (c) $MgSO_4$ is measured using AlGaN:Er^{3+} nanothermometer and plotted as a function of excitation laser intensity as shown in Fig. 5.1.

From Fig. 5.1, in all plots the local temperature change increased linearly with laser intensity. The decrease in slope for the local temperature change going from air to water is due to the additional heat dissipation channel available for nanowire excitation under water, as illustrated in the cartoon shown in the bottom right corner of Fig. 5.1. A further decrease in the local temperature slope is observed when ions are added to the pure water solution. Figure 5.1a–c shows that the largest change in

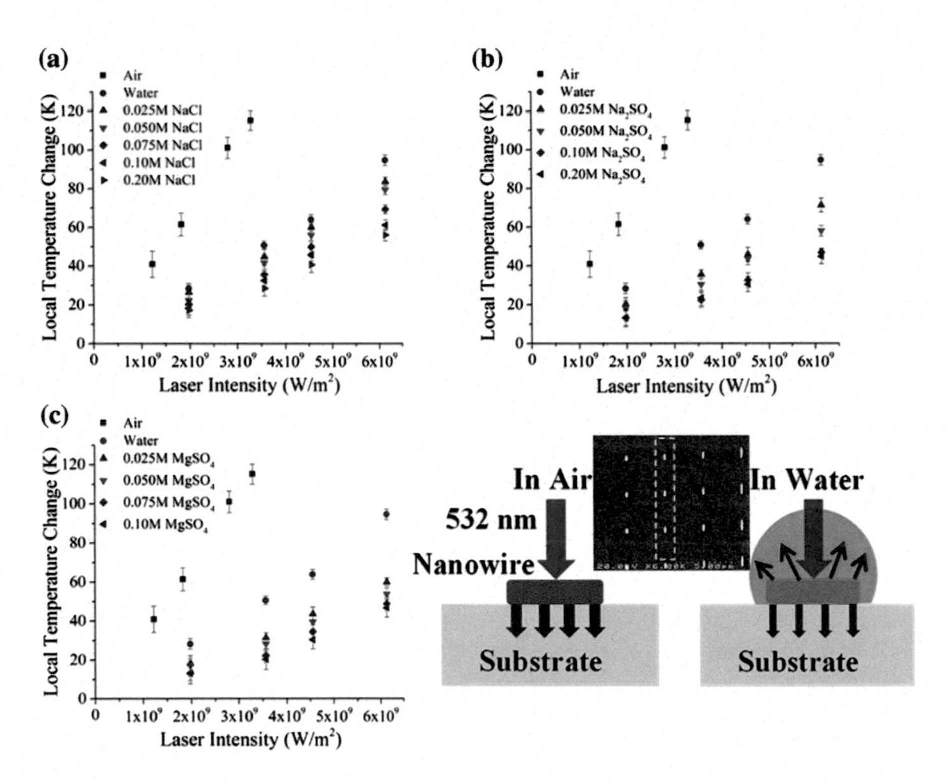

Fig. 5.1 A plot of local temperature change in K as a function of excitation laser intensity in W/m^2 for nanowire excitation under different concentration solutions of NaCl, Na_2SO_4 and $MgSO_4$. Cartoon shows a model of heat transfer with SEM image of the gold nanowires as inset. Error bars represents the uncertainty of temperature measurement. Reprinted with permission from *ACS Nano* 2016, *10*(6), 6080–6089. Copyright 2016 American Chemical Society

slope is observed for the salt with the largest charges, $MgSO_4$ shown in Fig. 5.1c. These results suggest that the charge of the salt in the solution is important to our measurement of local temperature change. All optical measurements on air, water and various ionic solutions were performed on the same nanowire, on the same day, to eliminate the possible error arising from the difference in cross sectional areas of the different Au NWs. Also, after performing all measurements in different solutions, same nanowire was again excited in pure water and the slope of the temperature change versus laser intensity plot was found/confirmed to be similar with that obtained in pure water before excitation under different solutions. Multiple measurements were also performed to ensure the reproducibility of the measurement using different nanowires and the similar trend was observed.

The slopes of temperature change versus laser intensity plots shown in Fig. 5.1 are plotted as a function of concentration of respective solution in molarity as shown in Fig. 5.2a, which reveals that the slopes decrease with increasing ion concentration. However, there is little agreement in the slopes for the different ions used.

The ion solution concentrations can be represented in terms of ionic strength using $I = \frac{1}{2}\sum_{i=1}^{n} c_i z_i^2$ where c_i is the molar concentration for each ion, and z_i is the charge number of that ion. Figure 5.2b shows the slopes as a function of ionic strength. Plotting the slopes as a function of ionic strength places all the data shown in Fig. 5.2a on a universal trend where slope decreases with ionic strength and approaches a saturation value. Any additional increase in ion concentration after ionic strength values of 0.3 results in very little or no changes in the value of the slope.

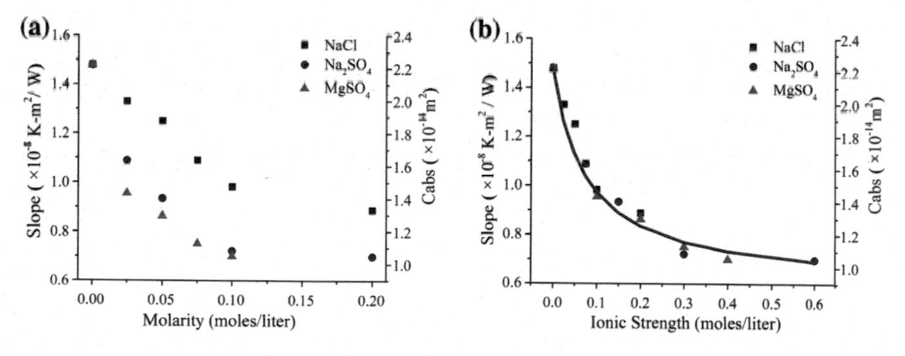

Fig. 5.2 A plot of slopes of temperature change versus laser intensity as a function of concentration of respective solution in **a** molarity and **b** ionic strength. Reprinted with permission from *ACS Nano* 2016, *10*(6), 6080–6089. Copyright 2016 American Chemical Society

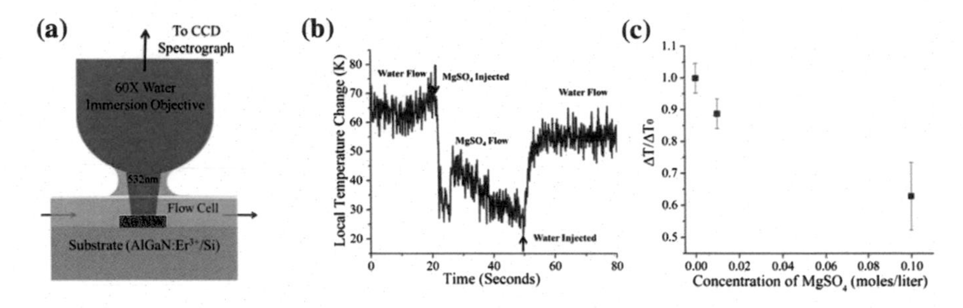

Fig. 5.3 **a** Schematic of the optical setup. **b** Temperature-time spectrum of a single optically excited gold nanowire in flow cell under the flow of pure water and 0.1 M $MgSO_4$ solution. **c** Relative drop in temperature during the flow of $MgSO_4$ solution with respect to the temperature change in pure water for 0.01 and 0.1 M $MgSO_4$ solution. Reprinted with permission from *ACS Nano* 2016, *10*(6), 6080–6089. Copyright 2016 American Chemical Society

5.3 Dynamic Temperature Measurements on Single Gold Nanowire Using Flow Cell

The temperature drop observed in Fig. 5.1 is further probed by performing dynamic measurements where the analyte could be changed at different intervals, while the temperature change on a single nanowire at 532 nm CW excitation being monitored continuously. Figure 5.3b shows the temperature-time spectrum of a single optically excited nanowire in flow cell under the flow of pure water and 0.1 M $MgSO_4$ solution. Arrows represent the injection of different fluids. As seen in figure, a relatively constant temperature change obtained throughout the flow of pure water dropped immediately after the injection of 0.1 M $MgSO_4$. Also, the temperature change increased and labelled back close to an initial value once pure water is injected and flowed through the channel while stopping the flow of $MgSO_4$ solution. This represents the extremely fast, sensitive and reversible influence of charges on the amount of heat generated from an optically excited single gold nanowire. Figure 5.3c shows the relative drop in temperature during the flow of $MgSO_4$ solution with respect to the temperature change in pure water for 0.01 and 0.1 M $MgSO_4$ solution. The result shows the sensitivity of the effect of charges on millimolar concentration range.

5.4 Model of Heat Transfer

In our measurements, we excite the same nanowire at similar laser intensities under pure water and different concentration aqueous solutions of NaCl, Na_2SO_4, and $MgSO_4$. The measured temperature changes (ΔT) are then plotted against corresponding laser intensities (I) as shown in Fig. 5.1. The heat dissipation from the optically heated nanowires into a spherical symmetric medium has been calculated

[4–6]. Under this approximation, the temperature change due to optical excitation is given by $\Delta T_{local} = \frac{C_{abs} I}{4\pi k_{eff} R_{eff}}$ where C_{abs} is the absorption cross section, I is the laser intensity, $k_{eff} = \frac{k_{sub} + k_{water}}{2}$ is the effective thermal conductivity and R_{eff} is the effective radius for a particle of volume equal to that of a sphere. As seen from Fig. 5.1, in all plots the local temperature change increased linearly with laser intensity. The slopes of these plots are then determined by linear regression using a zero intercept. This limit expects no temperature change when the laser does not heat the nanostructure. The slopes of the local temperature change versus laser intensity plots, therefore, are defined as $\left(\frac{\Delta T_{local}}{I}\right) = \left(\frac{C_{abs}}{4\pi k_{eff} R_{eff}}\right)$ [1, 7]. As the steady temperature is being measured on the same nanowire at similar laser intensities in different aqueous solutions, the changes on the measured slope, therefore, infers the changes on the amount of steady state heat on the nanowire. A decrease in slope of the local temperature change versus laser intensity reflects either a decrease in the amount of heat generated under optical excitation of nanowire (decrease in C_{abs}) under ionic solutions or enhanced heat dissipation from the nanowire into the surrounding fluid (increase in $G_{interface}$) because of the presence of ions at an interface.

The rate of heat dissipation depends upon the thermal parameters of the surrounding medium. For an optically-excited nanowire sitting on a substrate and immersed in air, the heat only dissipates into the substrate as the thermal conductivity of air is extremely low ($\sim$0.024 W/m-K). But for a nanowire immersed in water and aqueous solutions, heat dissipates into both the substrate and the surrounding water. Since the nanowires are lithographically fabricated onto the substrate, the rate of heat dissipation from the nanowire to the substrate does not depend upon the properties of the solution and should remain constant. Also, the concentration of the ionic solutions used in these measurements is relatively small (less than 0.2 M). At these low concentrations, the thermal conductivity and specific heat change by less than 0.2 and 4% respectively [8]. Consequently, the changes in the bulk properties of the solution cannot explain the large drop on the slopes of temperature change versus laser intensity plots.

5.5 Absorption Measurements on Gold Nanoparticle(s)/ Gold Nanorod(s)

The lithographic nanowires used for the measurements shown in Figs. 5.1 and 5.2 are patterned on non-transparent Silicon substrate, so absorption measurements were not feasible on these structures. A series of representative absorption measurements were, therefore, performed on single gold nanoparticles/nanorods to understand the mechanism of the drops in slopes of temperature change versus laser intensity plots and the temperature drop in dynamic measurements shown in Fig. 5.3. Single particle(s) absorption measurements were performed by bringing

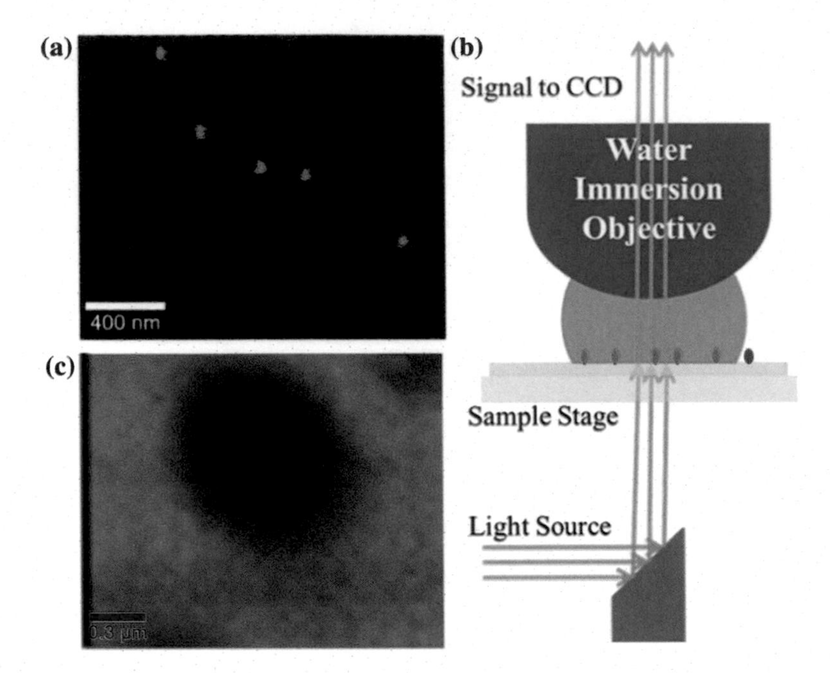

Fig. 5.4 **a** AFM image of single 40 nm gold nanoparticles spin coated onto the glass cover slip substrate. **b** Schematic optical setup for single particle(s) absorption measurements on gold nanoparticles/nanorods spin-coated onto the glass cover slip substrate. **c** Absorption image of a gold nanoparticles spin coated onto glass cover slip substrate.

the white light from the bottom part of the WITec α-SNOM300s microscope and collecting the transmitted light from the top part of the microscope through the water immersion objective and 25 μm collection fiber coupled with CCD spectrograph (Fig. 5.4).

Absorption measurement was first performed on 40 nm gold nanoparticle(s) under pure water droplet, followed by the measurement on the same nanoparticle(s) under droplet of an aqueous solution of 0.1 M $MgSO_4$. Absorbance of the nanoparticle for each measurement was then determined from the collected absorption images. Absorbance image was then generated by calculating the absorbance in each pixel location of the absorption image using the relation $A = -\log_{10}\frac{I}{I_0}$ where I and I_0 represent the spectrum intensity at the nanoparticle location and the background respectively. Numerous measurements were performed on single gold nanoparticle(s) and gold nanorods (100 nm × 25 nm) to check the reproducibility of the absorption measurements.

Figure 5.5a–c shows absorption spectra of 40 nm gold nanoparticles spin coated onto the glass cover slip substrate under pure water and 0.1 M $MgSO_4$ solution. Figure 5.5c, the processed absorption spectrum of a gold nanoparticle(s) does not show noticeable shift on the plasmon resonance wavelength but shows clear

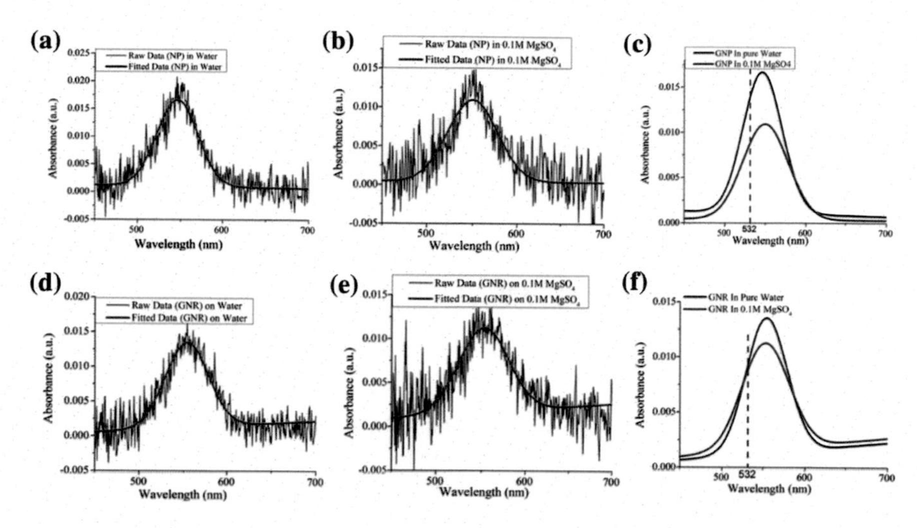

Fig. 5.5 Absorption spectra of **a–c** 40 nm gold nanoparticle(s), **d–f** 100 nm × 25 nm gold nanorods spin coated onto the glass cover slip substrate under pure water and 0.1 M $MgSO_4$ solution. The fitted spectra (**c** and **f**) do not show noticeable shift on the plasmon resonance wavelength but shows clear attenuation on plasmon absorbance of the gold nanoparticle(s) and nanorods immersed under ionic solution of $MgSO_4$. Reprinted with permission from *ACS Nano* 2016, *10*(6), 6080–6089. Copyright 2016 American Chemical Society

attenuation on plasmon absorbance of the gold nanoparticle immersed under ionic solution of $MgSO_4$. It is also evident that the extent of attenuation (or decrease) on plasmon absorbance of the nanoparticle is highly dependent on the excitation wavelength. Maximum attenuation (or decrease) in absorbance is seen to occur at surface plasmon resonance wavelength. This infers that the largest temperature difference will be observed if the excitation wavelength and the maximum surface plasmon resonance wavelength coincide. Figure 5.5d–f shows absorption spectra of 100 nm × 25 nm gold nanorods (Nanopartz) spin coated onto the glass cover slip substrate under pure water and 0.1 M $MgSO_4$ solution. Figure 5.5f, the processed absorption spectrum of a gold nanorod(s) also shows the attenuation on plasmon absorbance under ionic solution of $MgSO_4$.

5.6 Absorption and Temperature Measurements on a Same Gold Nanoparticle(s)

An absorption as well as temperature measurement on a same gold nanoparticle(s) is was performed by spin coating 40 nm gold nanoparticles on the transparent Sapphire substrate with thermal sensor film. Figure 5.6 shows the thermal profile of the nanoparticle under 532 nm CW excitation in pure water (black profile) and 0.1 M $MgSO_4$ (red profile). Absorption and temperature measurement on a

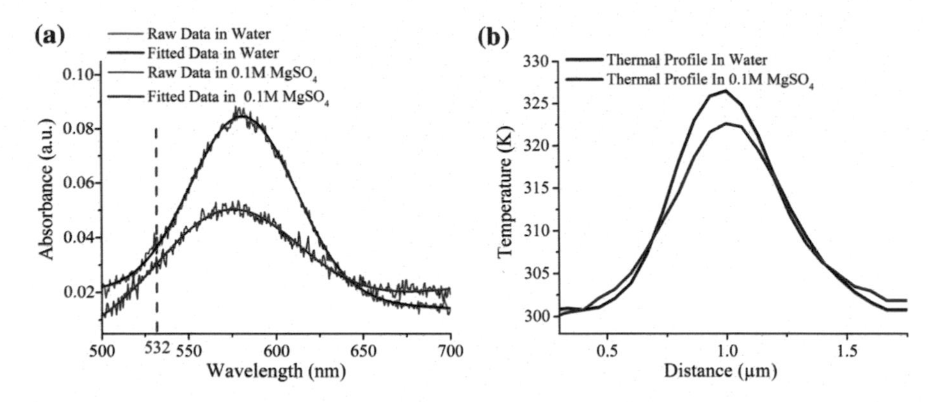

Fig. 5.6 **a** Absorption spectra and **b** the thermal profile under 532 nm CW excitation of 40 nm gold nanoparticle(s) spin-coated onto the substrate with thermal sensor film of AlGaN:Er^{3+} on sapphire glass. Reprinted with permission from *ACS Nano* 2016, *10*(6), 6080–6089. Copyright 2016 American Chemical Society

nanoparticle(s) was first performed in pure water followed by the absorption and temperature measurement on the same nanoparticle(s) in 0.1 M $MgSO_4$ solution. The temperature measurement in water and 0.1 M $MgSO_4$ was performed at the same intensity of 3×10^9 W/m^2. As seen from Fig. 5.6a, the absorption spectrum shows no noticeable shift in the plasmon absorption band when the solution is changed from pure water to 0.1 M $MgSO_4$ but shows attenuation of plasmon absorbance. Figure 5.6b shows the corresponding change in temperature when the solution is changed from pure water to 0.1 M $MgSO_4$. The thermal profile shows a decrease in maximum temperature change (ΔT_{max}) going from pure water to 0.1 M $MgSO_4$ solution. The absorption spectrum shows about 16% attenuation on plasmon absorbance going from pure water to 0.1 M $MgSO_4$ (from 0.039 in pure water to 0.033 in 0.1 M $MgSO_4$) and the corresponding drop in maximum temperature change (ΔT_{max}) is about 15% (drop from 327 in pure water to 323 in 0.1 M $MgSO_4$). The result suggests the temperature drop is a result of attenuation in the absorption when the solution is changed from pure water to 0.1 M $MgSO_4$. The temperature profiles (thermal images) of the nanoparticle(s) in pure water and 0.1 M $MgSO_4$ is shown in Fig. 5.6b.

5.7 Single Nanowire Dark-Field Scattering Measurements

Even though performing absorption measurements on the lithographic gold nanowires was feasible, dark-field scattering measurements were performed these nanowires. As absorption is directly related to the polarizability and scattering to polarizability squared, it is expected that changes in the scattering spectrum will be reflected as changes in the absorption spectrum. Figure 5.7 shows dark-field

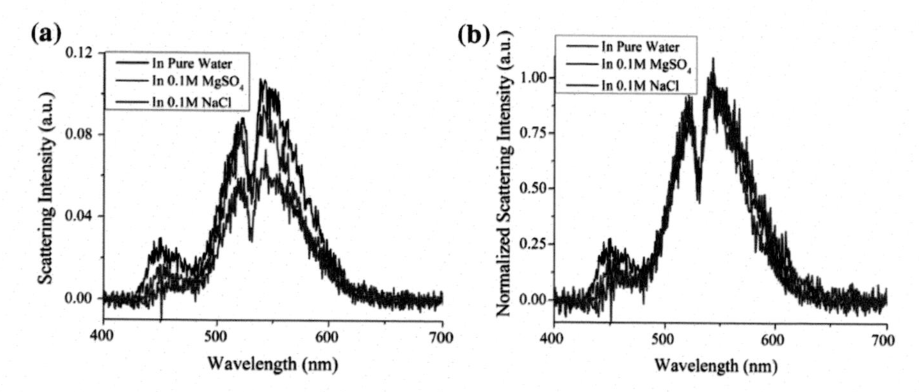

Fig. 5.7 **a** Dark-field scattering spectra of a single lithographic gold nanowire under pure water and 0.1 molar ionic solutions of NaCl and $MgSO_4$. The decrease (or attenuation) on net scattering intensity of the nanowire under ionic solutions of NaCl and $MgSO_4$ compared to that in pure water. **b** No noticeable shift on normalized scattering spectra of the nanowire under 0.1 M $MgSO_4$ and 0.1 M NaCl compared to that in pure water. Reprinted with permission from *ACS Nano* 2016, *10*(6), 6080–6089. Copyright 2016 American Chemical Society

scattering spectra of a single lithographic gold nanowire under pure water and 0.1 M ionic solutions of NaCl and $MgSO_4$. The scattering measurements shown in Fig. 5.7 are consistent with the results obtained from the absorption measurements.

5.8 Single Nanoparticle(s) Emission Measurements

Further investigation into the emission properties of the gold nanoparticles on glass substrates showed that the plasmon emission is also attenuated with solution ionic strength (Fig. 5.8). This result reinforces the assessment that the lower emission rate is due to an absorption attenuation because the lifetime and spontaneous emission rate are expected to be unaffected by the small change of ions in the solution.

5.9 Calculation of Absorption Cross Section of a Nanowire

The absorption cross section of the gold nanowire on a $Al_{0.94}Ga_{0.06}N$ film immersed in pure water and aqueous solutions can be calculated assuming that the nanowire has a uniform temperature and has an effective spherical structure with an effective radius of $R_{eff} = \left(\frac{3V}{4\pi}\right)^{1/3}$. This heat generation approximation from a point source with spherical symmetry presents little deviation for the correct value for our aspect ratio and nanowire size [9]. For a nanostructure where the heat dissipation can be characterized with spherical symmetry, the temperature change due to

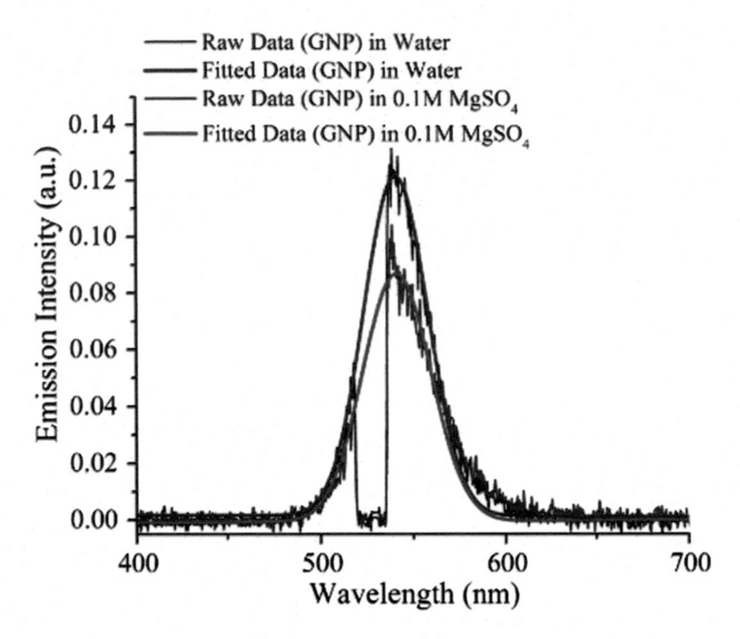

Fig. 5.8 Plasmon emission spectra at 532 nm CW excitation of 40 nm gold nanoparticle(s) spin-coated onto the glass cover slip substrate in pure water and 0.1 M $MgSO_4$ solution. Figure reveals the plasmon emission intensity is decreased for the nanowire excitation under ionic solution of 0.1 M $MgSO_4$. Reprinted with permission from *ACS Nano* 2016, *10*(6), 6080–6089. Copyright 2016 American Chemical Society

optical excitation is given by $\Delta T_{local} = \frac{C_{abs} I}{4\pi k_{eff} R_{eff}}$ where C_{abs} is the absorption cross section, I is the laser intensity, $k_{eff} = \frac{k_{sub} + k_{water}}{2}$ is the effective thermal conductivity and R_{eff} is the effective radius for a particle of volume equal to that of a sphere. The absorption cross section of a nanowire under optical excitation under pure water and various concentration solutions is then determined by rearranging the previous equation as $C_{abs} = 4\pi k_{eff} R_{eff} \left(\frac{\Delta T_{local}}{I}\right)$ where the quantity $\frac{\Delta T_{local}}{I}$ is given by the slope of temperature change versus laser intensity plot as shown in Fig. 5.1. Using the equation, an absorption cross section value of 2.25×10^{-14} m^2 is calculated for the nanowire excitation in pure water. The validity of the spherical approximation was examined and confirmed by the calculation of absorption cross section of the nanowire under the ellipsoid model with the equivalent radius and thermal-capacitance coefficient β as described in the literature [9]. The calculated values of the absorption cross for nanowire excitation under pure water and various concentration solutions of NaCl, Na_2SO_4 and $MgSO_4$ are plotted on right Y-axis in Fig. 5.2.

5.10 Langmuir Model of Charge Occupancy and Effect on Absorption Attenuation

Figure 5.2b shows that the absorption cross section of the gold nanowire changes with the solution ionic strength. Surface properties that depend upon ionic strength suggest that a Debye-Huckel screening length is important and that the electrical double layer between the gold surface and the surrounding solution is affecting the amount of light absorbed by the gold nanowire. The charge distribution at the interface changes with ionic strength. Figure 5.2b shows that the absorption cross section is a maximum when the ionic strength is zero, decreases as ionic strength is increased until the effect saturates at large ionic strength. This behavior suggests that accumulation of screened surface charge is responsible for the attenuation of the plasmon absorption. At a maximum occupancy, the effect saturates and increases in ionic strength does not further change the absorption cross section. This behavior can be modeled with a Langmuir adsorption model where ionic strength increases the screened charge occupancy (see Eq. 5.1). In this equation, C_o is the absorption cross at zero ionic strength, C_{sat} is the absorption cross section at saturation, C_{abs} is the absorption cross section at ionic strength μ, and K is an arbitrary constant defining the shape of the curve. The solid red line in Fig. 5.2b is the fit using Eq. 5.1 with K equal to 13.3.

$$\frac{C_o - C_{abs}}{C_o - C_{sat}} = \frac{K\mu}{1 + K\mu} \quad (5.1)$$

References

1. Green AJ, Alaulamie AA, Baral S, Richardson HH (2013) Ultrasensitive molecular detection using thermal conductance of a hydrophobic gold-water interface. Nano Lett 13(9):4142–4147
2. Srivastava SK, Gupta BD (2011) Influence of ions on the surface plasmon resonance spectrum of a fiber optic refractive index sensor. Sens Actuators B: Chem 156(2):559–562
3. Jain PK, Lee KS, El-Sayed IH, El-Sayed MA (2006) Calculated absorption and scattering properties of gold nanoparticles of different size, shape, and composition: applications in biological imaging and biomedicine. J Phys Chem B 110(14):7238–7248
4. Carlson MT, Green AJ, Richardson HH (2012) Superheating water by CW excitation of gold nanodots. Nano Lett 12(3):1534–1537
5. Govorov AO, Zhang W, Skeini T, Richardson H, Lee J, Kotov NA (2006) Gold nanoparticle ensembles as heaters and actuators: melting and collective plasmon resonances. Nanoscale Res Lett 1(1):84–90
6. Govorov AO, Richardson HH (2007) Generating heat with metal nanoparticles. Nano Today 2 (1):30–38

7. Carlson MT, Khan A, Richardson HH (2011) Local temperature determination of optically excited nanoparticles and nanodots. Nano Lett 11(3):1061–1069
8. Sharqawy MH, Lienhard JHV, Zubair SM (2010) Thermophysical properties of seawater: a review of existing correlations and data. Desalin Water Treat 16(1–3):354–380
9. Baffou G, Quidant R, de Abajo FJG (2010) Nanoscale control of optical heating in complex plasmonic systems. ACS Nano 4(2):709–716

Chapter 6
Photothermal Heating Study Using Er_2O_3 Photoluminescence Nanothermometry

Susil Baral, Ali Rafiei Miandashti and Hugh H. Richardson

6.1 Introduction

Measuring temperature at the nanoscale with high spatial (10 nm) and temperature (0.1 K) resolution remains challenging. Several methods have been recently developed to measure temperature at the nanoscale [1–5], but most of these techniques are either limited by spatial resolution or only infer temperature indirectly. A successful and intuitive approach is using sharp hybrid tip as a thermocouple to measure the nanoscale temperature [6, 7]. This approach can measure temperature with a spatial resolution at ~100 nm; however, this approach is extremely invasive due to tip punching of samples. More recently, several thermal microscopy techniques have been developed that are based on far-field optical measurements and consequently are less invasive [4, 5, 8–14], but the spatial resolution is either diffraction-limited ($\lambda/2NA$) or limited by the point spread function of the microscope. Near-field measurements can give much greater spatial resolution especially when used in conjunction with nanoparticle interactions [15, 16], but the local temperature is usually not measured directly. In this chapter, we introduce a new optical probe technique using a laser-trapped erbium oxide nanoparticle (~150 nm) that measures absolute temperature that is no longer limited by the point spread function of the microscope. We apply optical probe thermometry to measure the thermal profile away from an optically excited gold nanodot and confirm that the spatial resolution is limited by the cluster size and not limited by the point spread function of the microscope.

We name this technique as Scanning Optical Probe Thermometry (SOPT). Raster scanning is used to collect a thermal image of a gold nanodot. This temperature technique also depends on relative photoluminescence intensities of the $^2H_{11/2} \rightarrow {}^4I_{15/2}$ and the $^4S_{3/2} \rightarrow {}^4I_{15/2}$ energy transitions of the Er^{3+} ions. A convolution analysis of the thermal profile shows that the point spread function of our measurement is a Gaussian with a FWHM of 165 nm. We attribute the width of this function to clustering of Er_2O_3 nanoparticles in solution.

A. R. Miandashti et al., *Photo-Thermal Spectroscopy with Plasmonic and Rare-Earth Doped (Nano)Materials*, Nanoscience and Nanotechnology,
https://doi.org/10.1007/978-981-13-3591-4_6

6.2 Temperature Calibration of Erbium Oxide Photoluminescence

To evaluate the potential of nanoparticles for temperature measurements, Er_2O_3 nanoparticles photoluminescence emission was calibrated at different temperatures. The photoluminescence spectra from the peaks at 539 nm ($^2H_{11/2} \rightarrow {}^4I_{15/2}$ transition) and 564 nm ($^4S_{3/2} \rightarrow {}^4I_{15/2}$) were used to calibrate the temperature using $\frac{H}{S} = A\ \exp\left(-\frac{\Delta E}{kT}\right)$ where H is the peak area of the band at 539 nm and S is the peak area of the band at 564 nm. Plotting the natural log of $\frac{H}{S}$ versus reciprocal temperature yields a straight line (see Fig. 6.1) where the slope is equal to $(-\frac{\Delta E}{k})$ and the intercept is equal to the natural log of A (the pre-exponential term in the Boltzmann relationship).

The slope is equal to -1084 ± 65 K with an intercept of 1.9 ± 0.2. This gives a Boltzmann expression of $\left(\frac{H}{S}\right) = 6.68\mathrm{e}^{(-1084/T)}$ and an absolute temperature as $T = \frac{1084}{1.9 - \log_e(H/S)}$. The slope of -1084 K is related to the difference in energy between the H and S transitions [17] is within the uncertainty of the slope measurement The relative sensitivity ($S_{rel}(T) = \frac{1}{R}\frac{\partial R}{\partial T}$ where R is $\left(\frac{H}{S}\right)$) of the erbium

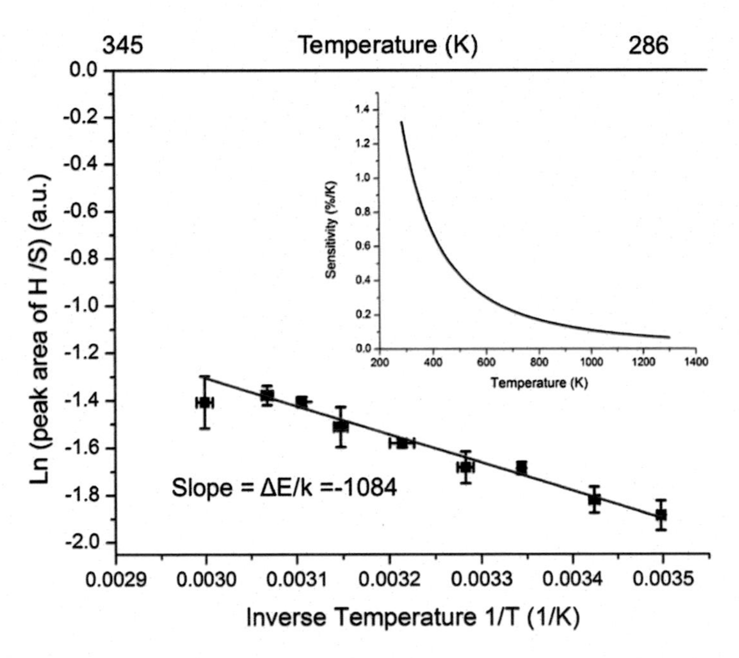

Fig. 6.1 Plot of natural log of the relative peak areas from the H (539 nm) and S (564 nm) bands with inverse temperature. The slope $\left(\frac{\Delta E}{k}\right)$ is -1084 ± 65 with an intercept of 1.9 ± 0.2 giving a Boltzmann expression of $\left(\frac{H}{S}\right) = 6.68\mathrm{e}(-1084/T)$. A plot of the sensitivity of Er^{3+} emission as a function of temperature is shown in the inset. Reprinted with permission from *Applied Physics A. (2016) 122: 340.* Copyright 2016 Springer

oxide nanoparticle sensor as a function of temperature is given in the figure insert. The relative sensitivity is given by $\frac{1084}{T^2}$ showing that the nanoparticle sensitivity decreases with temperature. At 1000 K the relative sensitivity has dropped by an order of magnitude. The temperature uncertainty at 300 K is ± 10 K (1 part in 30) if the data collection time per pixel is 0.03 s with a laser intensity of 3×10^{10} W/m^2. This uncertainty can be decreased if the integration time per pixel is increased and higher laser intensities are used. A temperature uncertainty of 2 K at 300 K will increase to a temperature uncertainty of 20 K at 1000 K.

Figure 6.2 shows a schematic diagram of the optical probe thermometry measurement. The erbium oxide nanoparticle ($\sim$150 nm) is trapped with the Nd:YAG laser (wavelength 532 nm, spot size $\sim$500 nm). Once the erbium oxide nanoparticle is trapped, it can be moved with the laser beam. The trapped particle emits light, and by moving the sample under the laser-trapped nanoparticle and collecting the luminescence, an image can be constructed.

The energy levels for erbium with typical photoluminescence spectra are shown in Fig. 6.3. 532 nm laser excites the Er_2O_3 nanoparticle photoluminescence emission from ($^2H_{11/2} \rightarrow {}^4I_{15/2}$, Labeled H) and ($^4S_{3/2} \rightarrow {}^4I_{15/2}$, Labeled S) transitions. The relative population of the different states is reflected in the relative peak intensities of H and S transitions. Temperature affects the population of the H and S states. Using the calibration curve described above the temperature of nanostructure can be determined.

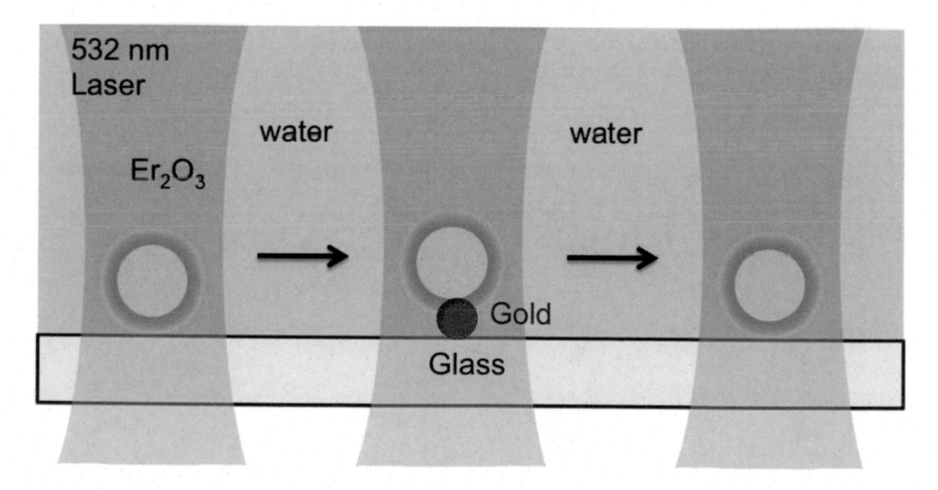

Fig. 6.2 Diagram of Optical Probe Thermometer. Er_2O_3 nanoparticle suspended in water is optically trapped with the 532 nm laser. The nanostructure (gold nanoparticle, nanowire or lithographically fabricated nanostructure) on glass cover slip is then scanned selectively under the trapped particle. The photoluminescence from the trapped Er_2O_3 nanoparticle is collected with the microscope objective and focused on the collection optical fiber. The emission is spectrally analyzed with a monochromator/CCD camera combination. Reprinted with permission from *Applied Physics A. (2016) 122: 340.* Copyright 2016 Springer

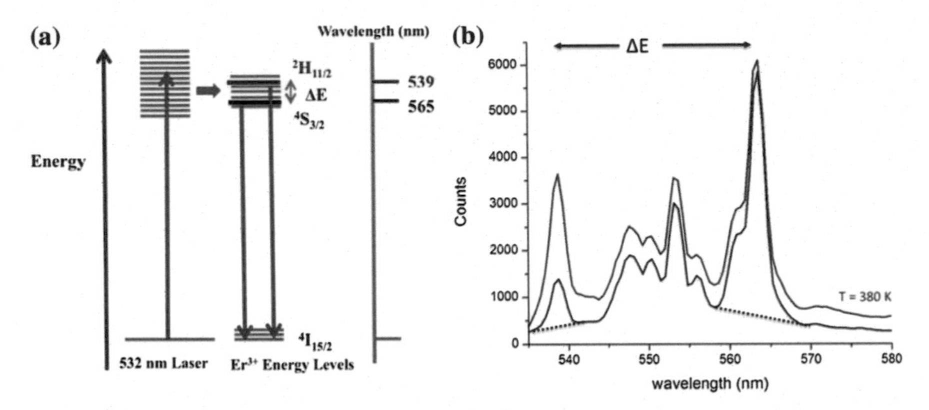

Fig. 6.3 **a** Energy level diagram of the Er_2O_3 nanoparticle. Er_2O_3 nanoparticles are excited with 532 nm light that populate $^2H_{11/2}$ and $^4S_{3/2}$ levels of Er^{3+}. The populations of these two states become thermalized resulting in photoluminescence intensities that are controlled by the relative populations given by a Boltzmann expression. The absolute temperature is determined by using this relationship knowing the difference in energy between the emitting states. **b** Photoluminescence spectrum of laser trapped Er_2O_3 nanoparticle at 300 K (black) and at 380 K (red). Reprinted with permission from *Applied Physics A. (2016) 122: 340.* Copyright 2016 Springer

6.3 Temperature Profile of Single Gold Nanodot

Figure 6.4a show a thermal image of a single gold nanodot collected by scanning optical probe thermometry (SOPT) method. Scanning a laser trapped optical probe (Erbium oxide nanoparticle) over a solid/water interface generates a temperature

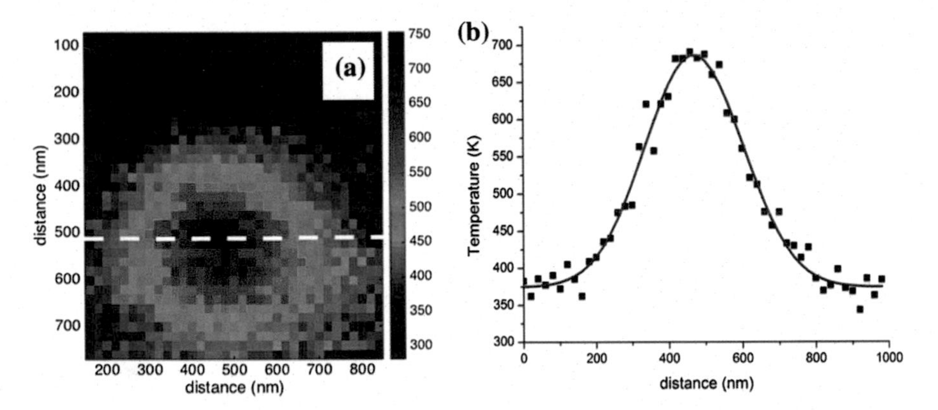

Fig. 6.4 **a** Thermal image of a nanostructure with the laser trapped Er_2O_3 nanoparticle. The white dotted line represents the place where the temperature profile shown in figure **b** is located. The solid red line is a Gaussian fit (FWHM of 300 nm) to the temperature profile. The temperature profile is taken in the direction of the scanned particle. The temperature image is collected as a horizontal scan that moves down vertically. Reprinted with permission from *Applied Physics A. (2016) 122: 340.* Copyright 2016 Springer

image. The interface is gold nanostructures on glass with a liquid phase (water). The absolute temperature is determined (see temperature calibration of Erbium oxide photoluminescence) from the erbium oxide nanoparticle photoluminescence spectra.

The image shown in Fig. 6.4 is generated from the scan on a 1 μm × 1 μm area. A typical imaging with scan parameters of 50 lines per image, 50 points per line, and an integration time of 0.05 s takes total time duration of 3 min and 3 s. The white dotted line on the image shows the location where a temperature profile is selected (shown in Fig. 6.4b). A Gaussian fit to the temperature profile is shown as the solid red line. The Gaussian has a full width at half maximum (FWHM) of 300 nm. This FWHM is narrower than the point spread function of the microscope (FWHM ~ 1500 nm) using the same collection parameters (100 μm collection fiber, 500 nm laser spot size) [8].

Figure 6.5 shows the comparison between the thermal images collected from several optical fibers compared to the thermal image collected by SOPT method using optically trapped Er_2O_3. a–c represents the dark field-scattering images of the lithographically fabricated nanodot (100 nm) in water generated using 100, 50 and

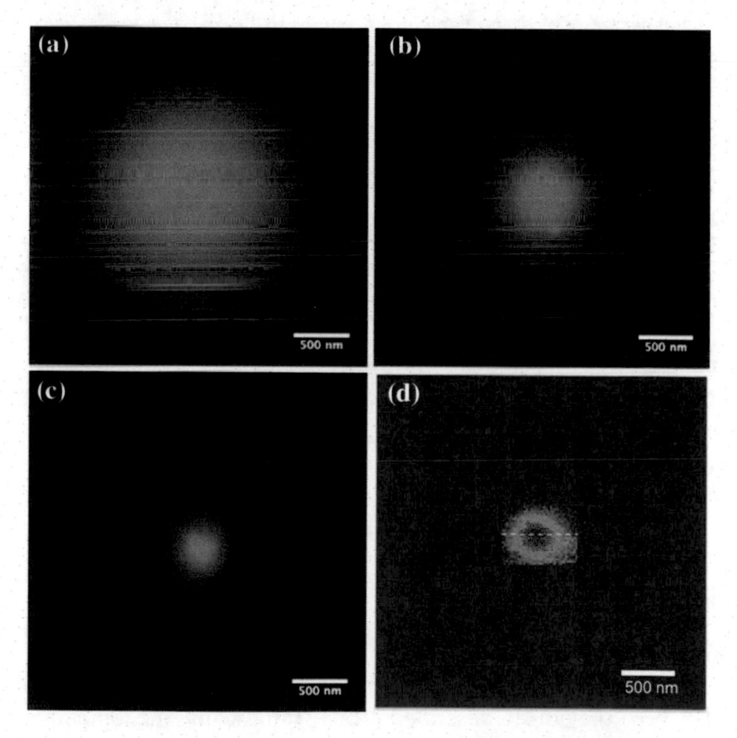

Fig. 6.5 **a–c** represents the dark field-scattering images of the lithographically fabricated nanodot (100 nm) in water collected using 100, 50 and 25 μm fiber respectively. **d** represents the comparative thermal image of the similar sized object (100 nm nanodot) collected using our technique (optically trapped Er_2O_3 nanoparticle) with 100 μm collection fiber. Reprinted with permission from *Applied Physics A. (2016) 122: 340.* Copyright 2016 Springer

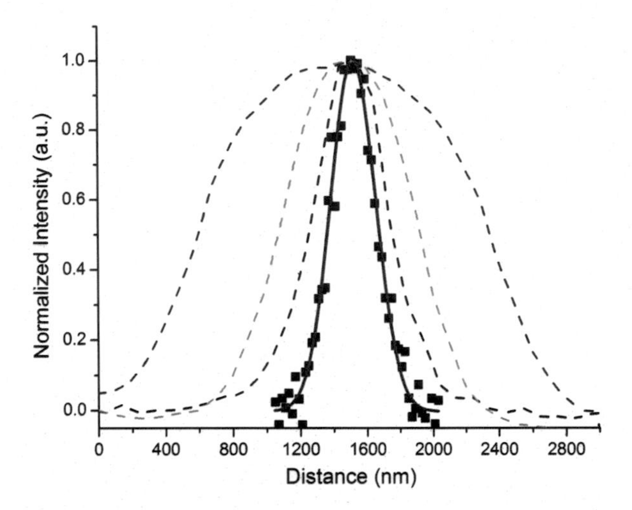

Fig. 6.6 Profile of the images shown in Fig. 6.5. The dotted purple dotted cyan and dotted black lines show the profile of an image collected using 100, 50 and 25 μm collection fiber respectively. The solid black line represents the profile of an image collected using SOPT technique with 100 μm collection fiber. Since our technique is not limited by the PSF due to the collection fiber, dramatic spatial resolution is achieved. Reprinted with permission from *Applied Physics A. (2016) 122: 340.* Copyright 2016 Springer

25 μm fibers respectively. Figure d represents the comparative thermal image of the similar sized object (100 nm nanodot) collected using our technique (optically trapped Er_2O_3 nanoparticle) with 100 μm collection fiber. Images a, b and c clearly show the spot size and hence the point spread function (PSF) of the microscope depends on the collection fiber. Image d shows that our technique is no longer limited by the PSF of our collection fiber as seen from images a, b and c but is limited by the size of the Er_2O_3 nanoparticle trapped for imaging. This signifies the ability of our technique to achieve sub-diffraction thermal imaging and temperature measurements. Figure 6.6 indicates the dotted purple, cyan and black lines showing the profile of an image collected using 100, 50 and 25 μm collection fiber respectively. The solid black line represents the profile of an image collected using SOPT technique with 100 μm collection fiber. Since the technique is not limited by the PSF due to the collection fiber, a marked spatial resolution is achieved.

Figure 6.7 shows the convolution graphs of thermal profile of experimental data from the theoretical plots. The experimental temperature profile shown in Fig. 6.7c is a result of a convolution with the point spread function of the scanning optical probe thermometer measurement and the thermal profile. The theoretical thermal profile from an optically heated gold nanodot assuming no interface thermal resistance has been previously described [16, 18]. Basically, the temperature at the nanodot should be constant, and the temperature should decrease as 1/r away from the nanodot. This true temperature profile will be slightly different than the expected profile because we are moving the nanoheater under our laser source and measuring the temperature at the maximum laser position. This scanning of the

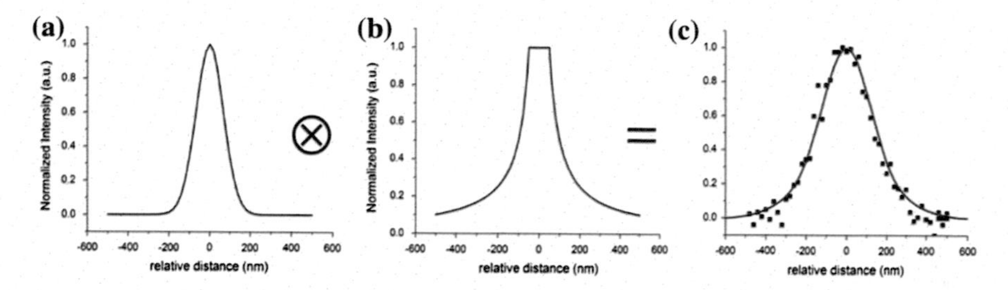

Fig. 6.7 **a** A Gaussian function having a FWHM of 165 nm convoluted with the theoretical temperature profile from a 100 nm diameter optically heated nanodot (panel **b**) to give the experimental temperature profile (panel **c**). The solid red line is the convolution of **a** with **b**. The convolution is superimposed on the experimental profile (solid black squares). All curves have been normalized and plotted in relative distance around the peak maximum. Reprinted with permission from *Applied Physics A. (2016) 122: 340.* Copyright 2016 Springer

nanoheater will clip the wings of the expected 1/r profile. This introduces a small perturbation to the expected profile shown in Fig. 6.5c, but this small perturbation will not significantly affect the convolution. The temperature has been normalized, and the distance has been centered on the nanodot. The Gaussian profile that gives the best convolution with the experimental profile is shown in Fig. 6.7a. The FWHM of the Gaussian profile is 165 nm. Figure 6.7c shows the convolution (shown as the solid red line) superimposed on the normalized experimental profile. These results suggest that the point spread function of the scanning optical probe thermometer is around 165 nm. This is larger than the average particle size of the erbium oxide nanoparticles (45 nm), indicating that there is some clustering of particles in solution. TEM image of samples made from dried erbium oxide nanoparticle solutions confirm that clustering in solution is occurring (see Fig. 6.8). The size and amount of clustering is affected by the amount of sonication as well as the type and amount of surfactant.

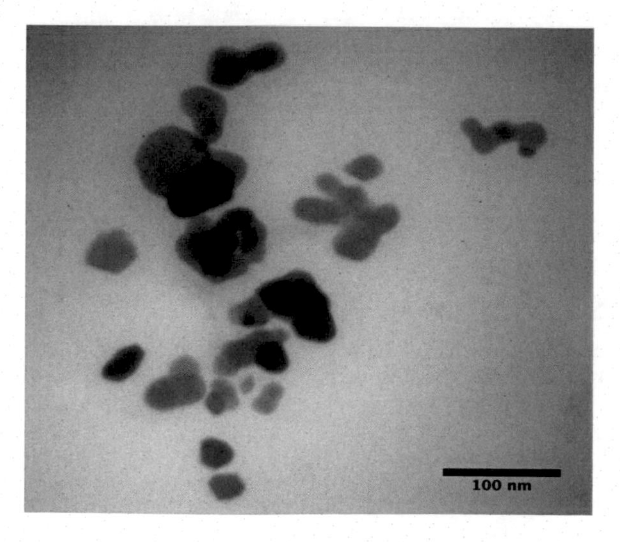

Fig. 6.8 TEM image of erbium oxide nanoparticle clustering in solution

6.4 Temperature Measurement Inside a Microbubble

One of the advantages of SOPT technique is the application of optical thermometery in heat transfer studies. Optically trapped Er_2O_3 nanoparticle can be placed to a desired location and the temperature of your system can be studied. In one example, we used this technique to measure the bubble nucleation temperature. The vapor nucleation temperature for a cluster of 100 nm × 25 nm gold nanorods immobilized on glass and immersed in degassed water was studied with the optical probe thermometer. In this measurement, the Er_2O_3 nanoparticle is trapped and then moved to the gold nanorods. The gold nanorods and Er_2O_3 nanoparticle have a relatively strong interaction that keeps the nanoparticle in place during the duration of the measurement. The laser intensity is increased stepwise from the starting value of 8×10^9 W/m^2 until a nucleation event takes place. Figure 6.9 shows the temperature time spectrum for nanorod excitation with increasing laser intensity. A single nucleation event is observed in the time spectrum at 165 s with the laser intensity value of 4×10^{10} W/m^2. The temperature corresponding to vapor nucleation is at 555 K (the spinodal decomposition temperature of water). This nucleation temperature is consistent with a recent molecular dynamics simulation

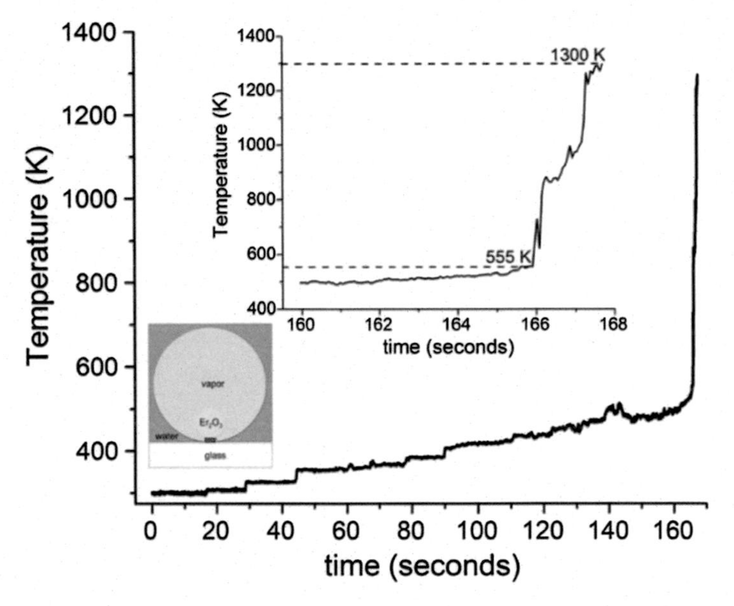

Fig. 6.9 Temperature-time spectrum of the gold nanorods immobilized on glass and immersed in degassed water. The laser intensity is increased stepwise starting from 8.12×10^9 W/m^2 to the value of 4.3×10^{10} W/m^2 at which bubble nucleation takes place. The region of the spectrum with the nucleation event is magnified and shown on top. The nucleation event is characterized by an abrupt increase in temperature and occurs at the temperature of 555 K. The small inset in lower left corner is a cartoon depicting the expected configuration of the nucleation bubble. Reprinted with permission from *Applied Physics A. (2016) 122: 340.* Copyright 2016 Springer

[19] and our previous measurements that heating a nanoscale object in degassed water is superheated to the spinodal decomposition temperature of water [20, 21].

The temperature of vapor nucleation using nanometer sized optically heated particles is expected to depend upon the volume of water heated beyond the boiling temperature [21]. Greatly reducing the volume of water at a temperature higher than the boiling point results in vapor nucleation occurring at the spinodal decomposition temperature (SDT) of water ($\sim$0.9 TC) [18, 21]. Vapor nucleation at a lower temperature than the SDT occurs when nucleation site the liquid helps the nucleation process. Heterogeneous nucleation dynamics is a statistical process that should depend exponentially on temperature [19, 22]. A single nucleation event is recorded in Fig. 6.9 that is near the spinodal decomposition temperature of water. The volume of heated water is greatly reduced with a single nanoheater resulting in a nucleation process that occurs homogeneously at a temperature near the SDT. The cartoon in the lower left-hand corner is the expected configuration of the nucleation bubble. A vapor cocoon encases the Er_2O_3 nanoparticle sensor with the optically excited nanoheater. The extreme temperature rise of the sensor and heater is due to the inhibition of heat transfer to the surrounding liquid by the vapor bubble. The temperature plateau at 1300 K is assigned to a temperature where the gold nanostructure melts.

6.5 Drawbacks/Limitations of the Technique

The efficiency of trapping is dependent on the dispersion of the Er_2O_3 nanoparticles and the size of the nanoparticle chosen for trapping. For the properly dispersed Er_2O_3 nanoparticles (which can be achieved by the addition of surfactant and dispersant), bigger cluster of particles is easy to trap and stable upon scanning. Since the spatial resolution of temperature measurement is dependent on the size of the Er_2O_3 nanoparticle, trapping the smaller particles is desired. But it is extremely difficult to trap a single Er_2O_3 nanoparticle with size of $\sim$40 nm. Since the Er_2O_3 particles tend to form clusters, smaller cluster of particles with the size $\sim$100–200 nm can be trapped well and are stable on scanning as well. Also, the relatively weak forces caused by laser trapping of small nanoparticles result in several limitations. Any strong interaction will usually override the weak forces in the laser trap. Instabilities associated with the trap are another concern for imaging which results in the particles being removed out of the trap during scan. Electrostatic interactions between surfaces and nanostructures on the surface with the Er_2O_3 nanoparticles usually override the weak trapping forces resulting in the inability to move the thermal sensor particles after falling into the electrostatic trap. Surfactant molecules surrounding the thermal sensor nanoparticle cluster help alleviate some of these problems. Also, the surface charge can be changed with monolayers of charged polymers. Brownian motion of the particle within the trap leads to the fluctuations on the spatial resolution leading to the temperature being averaged over the volume of fluctuation, and hence poorer spatial resolution.

References

1. Boyer D, Tamarat P, Maali A, Lounis B, Orrit M (2002) Photothermal imaging of nanometer-sized metal particles among scatterers. Science 297(5584):1160–1163
2. Cognet L, Tardin C, Boyer D, Choquet D, Tamarat P, Lounis B (2003) Single metallic nanoparticle imaging for protein detection in cells. Proc Natl Acad Sci USA 100(20):11350–11355
3. Berciaud S, Cognet L, Blab GA, Lounis B (2004) Photothermal heterodyne imaging of individual nonfluorescent nanoclusters and nanocrystals. Phys Rev Lett 93:(25)
4. Baffou G, Kreuzer MP, Kulzer F, Quidant R (2009) Temperature mapping near plasmonic nanostructures using fluorescence polarization anisotropy. Opt Express 17(5):3291–3298
5. Baffou G, Bon P, Savatier J, Polleux J, Zhu M, Merlin M, Rigneault H, Monneret S (2012) Thermal imaging of nanostructures by quantitative optical phase analysis. ACS Nano 6 (3):2452–2458
6. Pollock HM, Hammiche A (2001) Micro-thermal analysis: techniques and applications. J Phys D-Appl Phys 34(9):R23–R53
7. Sadat S, Tan A, Chua YJ, Reddy P (2010) Nanoscale thermometry using point contact thermocouples. Nano Lett 10(7):2613–2617
8. Carlson MT, Khan A, Richardson HH (2011) Local temperature determination of optically excited nanoparticles and nanodots. Nano Lett 11(3):1061–1069
9. Loew P, Kim B, Takama N, Bergaud C (2008) High-spatial-resolution surface-temperature mapping using fluorescent thermometry. Small 4(7):908–914
10. Vetrone F, Naccache R, Zamarron A, Juarranz de la Fuente A, Sanz-Rodriguez F, Martinez Maestro L, Martin Rodriguez E, Jaque D, Garcia Sole J, Capobianco JA (2010) Temperature sensing using fluorescent nanothermometers. ACS Nano 4(6):3254–3258
11. Li S, Zhang K, Yang J, Lin L, Yang H (2007) Single quantum dots as local temperature markers. Nano Lett
12. Van de Broek B, Grandjean D, Trekker J, Ye J, Verstreken K, Maes G, Borghs G, Nikitenko S, Lagae L, Bartic C, Temst K, Van Bael MJ (2011) Temperature determination of resonantly excited plasmonic branched gold nanoparticles by X-ray absorption spectroscopy. Small 7(17):2498–2506
13. Setoura K, Werner D, Hashimoto S (2012) Optical scattering spectral thermometry and refractometry of a single gold nanoparticle under CW laser excitation. J Phys Chem C 116 (29):15458–15466
14. Bendix PM, Nader S, Reihani S, Oddershede LB (2010) Direct measurements of heating by electromagnetically trapped gold nanoparticles on supported lipid bilayers. ACS Nano 4 (4):2256–2262
15. Lee J, Govorov AO, Kotov NA (2005) Bioconjugated superstructures of CdTe nanowires and nanoparticles: multistep cascade forster resonance energy transfer and energy channeling. Nano Lett 5(10):2063–2069
16. Govorov AO, Zhang W, Skeini T, Richardson H, Lee J, Kotov NA (2006) Gold nanoparticle ensembles as heaters and actuators: melting and collective plasmon resonances. Nanoscale Res Lett 1(1):84–90
17. Garter MJ, Steckl AJ (2002) Temperature behavior of visible and infrared electroluminescent devices fabricated on erbium-doped GaN. IEEE Trans Electron Device 49(1):48–54
18. Govorov AO, Richardson HH (2007) Generating heat with metal nanoparticles. Nano Today 2(1):30–38
19. Liang Z, Sasikumar K, Keblinski P (2014) Thermal transport across a substrate-thin-film interface: effects of film thickness and surface roughness. Phys Rev Lett 113:(6)
20. Carlson MT, Green AJ, Khan A, Richardson HH (2012) Optical measurement of thermal conductivity and absorption cross-section of gold nanowires. J Phys Chem C 116(15):8798–8803

21. Baral S, Green AJ, Livshits MY, Govorov AO, Richardson HH (2014) Comparison of vapor formation of water at the solid/water interface to colloidal solutions using optically excited gold nanostructures. ACS Nano 8(2):1439–1448
22. Kotaidis V, Plech A (2005) Cavitation dynamics on the nanoscale. Appl Phys Lett 87(21):3

Chapter 7
Nanoscale Temperature Study of Plasmonic Nanoparticles Using NaYF4:Yb^{3+}:Er^{3+} Upconverting Nanoparticles

Ali Rafiei Miandashti, Susil Baral and Hugh H. Richardson

7.1 Introduction

Nanocrystals containing Er^{3+} ions are one of the most widely used nanoparticles for measuring the photothermal heating from plasmonic nanoparticles [1–8]. $NaYF_4$:Yb^{3+},Er^{3+} nanocrystals emit a green luminescence emission when irradiated with near-IR light. The green band emission of Er^{3+} doped nanocrystals are used for luminescence thermometry [3, 5, 6]. However, One of the significant effects of gold nanoparticles on luminescence emission of lanthanide ions in crystalline structure is enhancement and quenching phenomena [9]. The interaction of gold nanoparticles with UCNPs leads to either increasing the number of energy transfers to excited states or reducing the number of rapid surface related non-radiative processes [10, 11]. Since Er^{3+} ion luminescence thermometry is growing, there is a need to confirm if the temperature dependent quenching rate is different between the H ($^2H_{11/2} \rightarrow {}^4I_{15/2}$ transition) and S ($^4S_{3/2} \rightarrow {}^4I_{15/2}$ transition) bands. We measured the temperature dependence in the steady-state emission of the H and S bands and compared the calculated temperature to the measured temperature to confirm that these UCNPs decorated with gold nanoparticles act as valid thermal sensors.

7.2 Temperature Calibration of $NaYF_4$:Yb^{3+},Er^{3+} Nanocrystals Photoluminescence

In order to use $NaYF_4$:Yb^{3+},Er^{3+} nanocrystals for measurement of temperature at the nanoscale we generated the calibration curve. The green photoluminescence emission of $NaYF_4$:Yb^{3+},Er^{3+} nanocrystals is temperature dependent. The luminescence emission of upconverting nanoparticles is shown in Fig. 7.1. The emission spectrum shows four different bands in blue, green and red regions.

A. R. Miandashti et al., *Photo-Thermal Spectroscopy with Plasmonic and Rare-Earth Doped (Nano)Materials*, Nanoscience and Nanotechnology,
https://doi.org/10.1007/978-981-13-3591-4_7

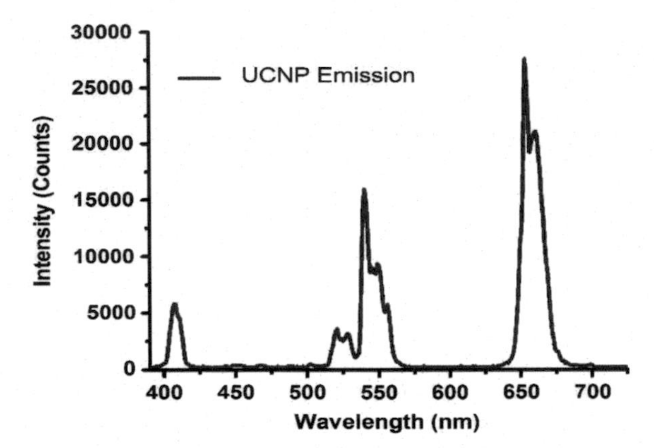

Fig. 7.1 Luminescence emission of $NaYF_4:Yb^{3+}:Er^{3+}$ nanocrystals showing four emission bands in blue, green and red regions

The emission bands in green region, which are $^2H_{11/2} \rightarrow {}^4I_{15/2}$ and $^4S_{3/2} \rightarrow {}^4I_{15/2}$ bands, are thermally coupled and the ratio of the peaks depend on temperature.

Figure 7.2a shows the luminescence spectrum in the green region, known as H and S bands. Calculation of temperature based on luminescence emission has been previously introduced [8]. In this approach, we used the green band emissions (H and S bands) to calculate temperature for UCNPs and UCNP/GNPs under different temperatures. The temperature of the nanoparticles was changed using a Peltier where the temperature changed from 0 to 100 °C and it was monitored by a thermocouple. As Fig. 7.2a shows, at higher temperatures we observe an increase in the intensity of H band and a decrease in the intensity of S band. Therefore, the ratio of H band to S band can be used as criteria to calculate the temperature. We observed a small difference in the S band peak shape with temperature that is attributed to a slight broadening of the peaks at higher temperatures and underlying

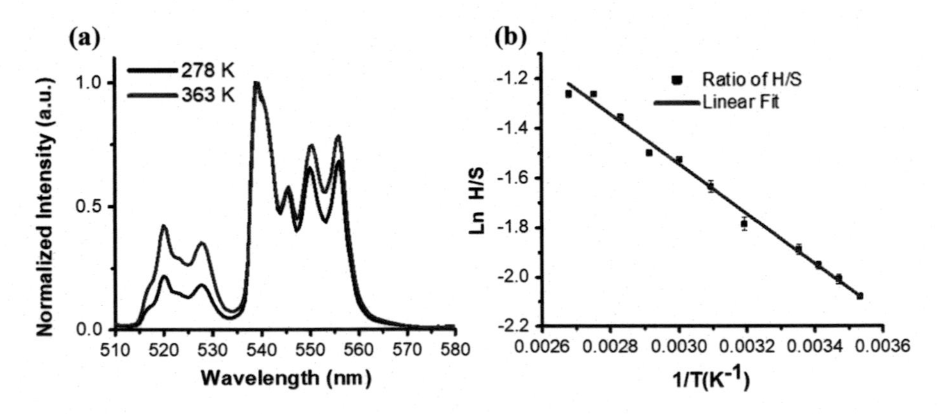

Fig. 7.2 **a** Green band emission of UCNPs under 980 nm laser illumination at two different temperatures. **b** Calibration plot of UCNP emission between the ranges of 0–90 °C. Reprinted (adapted) with permission from *ACS Photonics, 2017, 4(7), pp 1864–1869*. Copyright 2017 American Chemical Society

peaks that do not change with temperature. By having the ratio of H band to S band and the Boltzmann's equation $\left(\frac{H}{S} = A\,exp\left(\frac{-\Delta E}{kT}\right)\right)$ where A is the pre-exponential factor and ΔE is an energy difference between the H band ($^2H_{11/2} \rightarrow {}^4I_{15/2}$ transition) and the S band ($^4S_{3/2} \rightarrow {}^4I_{15/2}$ transition), and k as Boltzmann's constant, we calculated the temperature. The pre-exponential factor, A is defined from fitting the best linear fit to the scattering plot of natural logarithm of H/S versus 1/T. By taking the relative peak areas of H and S bands at various temperatures, we can plot the calibration curve and use it for our temperature measurements, Fig. 7.2b shows a slope of 996 ± 65 and pre-exponential factor of 1.4 ± 0.2 for the rest of temperature calculations. The uncertainty in calculation of temperature, ±8.5 K is due to the fluctuations of peak intensities and it can be improved by increasing the integration time of data collection.

7.3 Characterization of $NaYF_4:Yb^{3+},Er^{3+}$ Nanocrystals

Figure 7.3a shows the TEM image of $NaYF_4:Yb^{3+},Er^{3+}$ nanocrystals synthesized through thermal decomposition method [12]. $NaYF_4:Yb^{3+},Er^{3+}$ nanocrystals have average diameter of 300 nm and coated with hydrophobic layer of oleic acid as capping agent. In order to attach gold nanoparticles to the surface of $NaYF4:Yb^{3+}$, Er^{3+} nanocrystals, oleic acid is removed from the surface as reported in the literature [13]. $NaYF_4:Yb^{3+},Er^{3+}$ nanocrystals were attached to 10 nm gold nanoparticles through an electrostatic interaction (Fig. 7.3b). Figure 7.3c and d show the high resolution TEM image of a gold nanoparticle attached onto a $NaYF4:Yb^{3+}$, Er^{3+} nanocrystal with different magnifications.

The photothermal response of decorated upconverting nanoparticles has to be evaluated. Figure 7.4a shows the response of $NaYF_4:Yb^{3+},Er^{3+}$ nanocrystals and $NaYF_4:Yb^{3+},Er^{3+}$ nanocrystals decorated with gold nanoparticle to the intensity of 980 nm laser. A slight increase in temperature with 980 nm light is observed for the UCNPs (~10 K) while a large temperature increase is observed for the UCNP/GNPs (~150 K). The linear increase in temperature for the UCNP/GNPs is due to proximity of clusters of UCNP on the surface of cover slip. When a film of UCNP/GNPs are formed, the gold nanoparticles interaction causes a change in the plasmonic bands of gold that shift toward 980 nm leading to an increase in absorption at 980 nm and subsequently an increase in temperature (Fig. 7.4a).

The Jablonski energy level diagram for Er^{+3} and Yb^{+3} ions are shown in Fig. 7.4b. Upconversion occurs because the sensitizer (Yb^{3+}) absorbs light at 980 nm and transfers electrons to Er^{3+} where a second photon is absorbed to promote the population of electron to the $^2H_{11/2} \rightarrow {}^4I_{15/2}$ and $^4S_{3/2} \rightarrow {}^4I_{15/2}$ levels.

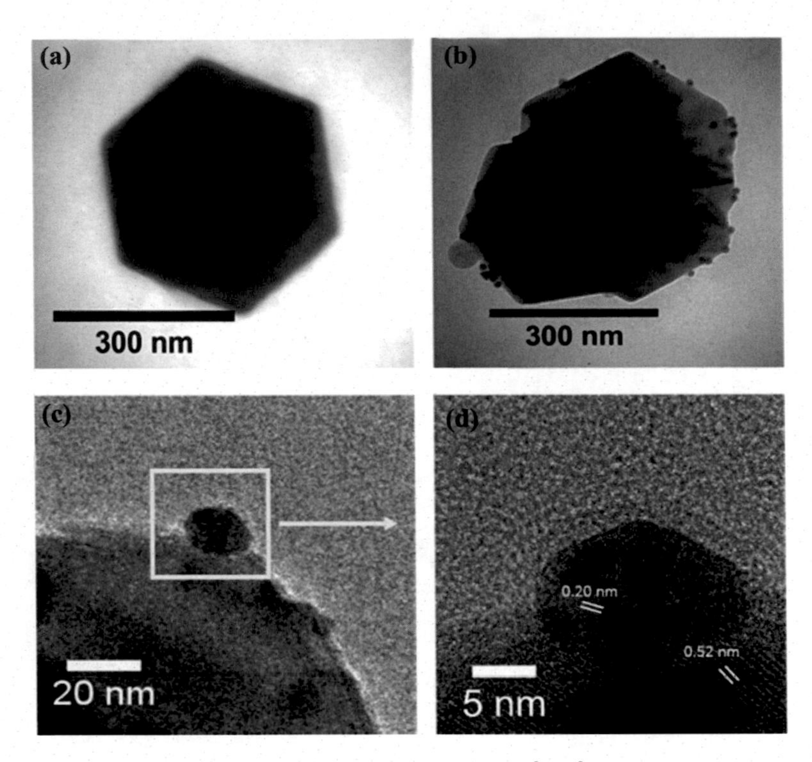

Fig. 7.3 **a** TEM image of β-phase hexagonal $NaYF_4$:Yb^{3+},Er^{3+} upconverting nanoparticle. **b** TEM image of hexagonal $NaYF_4$:Yb^{3+},Er^{3+} upconverting nanoparticles decorated with gold nanoparticles. **c** and **d** High-resolution TEM image of gold nanoparticle attached to $NaYF_4$:Yb^{3+}, Er^{3+} nanocrystal. Reprinted with permission from *ACS Photonics, 2017, 4(7), pp 1864–1869*. Copyright 2017 American Chemical Society

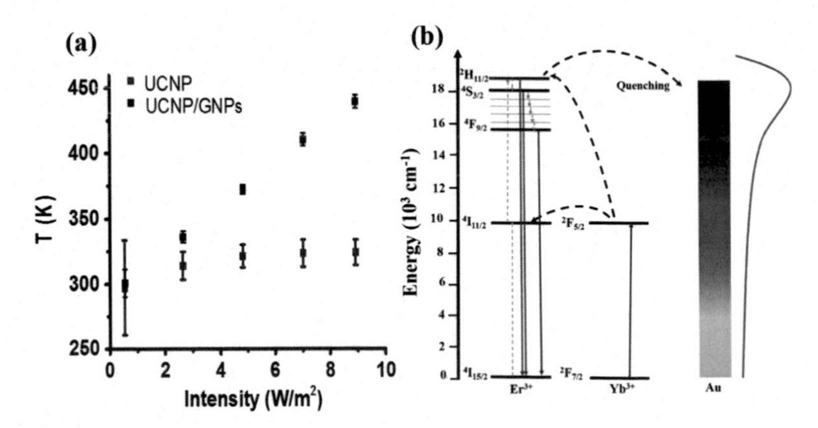

Fig. 7.4 **a** The temperature response of UCNPs and UCNP/GNPs to the intensity of 980 nm laser. **b** Jablonski energy diagram of UCNP/GNPs and possible quenching mechanism due to the presence of gold nanoparticles. Reprinted with permission from *ACS Photonics, 2017, 4(7), pp 1864–1869*. Copyright 2017 American Chemical Society

7.4 Lifetime Study of $NaYF_4$:Yb^{3+},Er^{3+} Nanocrystals

To understand the potential of UCNPs to measure the temperature of gold nanostructures specially at higher temperatures, we made time-resolved measurements of the green emission bands for UCNPs and UCNP/GNPs (Fig. 7.5a). Figure 7.5A shows the time-resolved emission from the $^2H_{11/2} \rightarrow {}^4I_{15/2}$ band ($\sim$520 nm) for UCNPs and UCNP/GNPs with pulsed 980 nm laser excitation. The blue dotted line shows the pulse profile of the excitation laser. The solid red and black lines show the time-resolved emission for UCNPs and UCNP/GNPs in response to the laser pulse respectively. We will keep the color-coding that red data is for the UCNPs and black data is from UCNP/GNPs. Figure 7.5a shows that the time to reach steady-state is nearly the same for the UCNP/GNPs and UCNPs. The seed nanoparticles do not appear to alter the time needed to reach steady-state. However, we observe that the amount of emission is less for the UCNPs/GNPs compared to the UCNPs without gold suggesting that the gold nanoparticles open up a non-radiative channel that quenches the Er^{3+} excited state emission. Supporting this hypothesis, we observe that after 980 nm light is turned off, the pathway for energy dissipation between the UCNP/GNPs and UCNPs is different.

A faster decay is observed for the UCNP/GNPs than for the UCNPs. A plot of the natural log of the emission versus time is shown in Fig. 7.5b for the UCNP/GNPs and UCNPs. The solid lines in the plot are fits to the data. The UCNPs have a single exponential fit while the UCNP/GNPs data is fitted with a double exponential. The lifetime for the UCNPs and UCNP/GNPs is compared in Fig. 7.5c. The lifetime of the UCNPs is $\sim$230 μs and the lifetimes for the UCNP/GNPs are 20 μs (weighting coefficient of $\sim$0.75) and 180 μs (weighting coefficient of $\sim$0.25). Table 7.1 gives a summary of the lifetimes and weighting coefficients for both the $^2H_{11/2} \rightarrow {}^4I_{15/2}$ ($\sim$520 nm) and the $^4S_{3/2} \rightarrow {}^4I_{15/2}$ band ($\sim$540 nm).

These NaYF4:Yb^{3+},Er^{3+} nanocrystals are used as temperature sensors by collecting the emission spectrum and evaluating the ratio of peak areas from the $^2H_{11/2} \rightarrow {}^4I_{15/2}$ and $^4S_{3/2} \rightarrow {}^4I_{15/2}$ bands. We observe a change in the lifetime of the UCNPs when gold seed particles are attached to the UCNPs and there appears to be differences in the lifetime between $^2H_{11/2} \rightarrow {}^4I_{15/2}$ and $^4S_{3/2} \rightarrow {}^4I_{15/2}$ bands when the UCNPs are decorated (see Table 7.1). This behavior led us to measure the temperature dependence in the lifetime for UCNP/GNPs and UCNPs to see if these UCNPs can be used as an effective temperature sensor. Figure 7.6a shows the temperature dependence in the lifetime for the H band ($^2H_{11/2} \rightarrow {}^4I_{15/2}$ transition, black squares) and S band ($^4S_{3/2} \rightarrow {}^4I_{15/2}$ transition, red circles) for UCNPs. The lifetime of these two bands decreases with temperature; suggesting that the emission from the H and S bands are quenched with temperature. This temperature dependence in the lifetime is also observed for UCNP/GNPs. This data shows that the lifetimes associated with the UCNP/GNPs, for both the H band and S band decrease with temperature.

A decrease in the lifetime with temperature suggests that there is quenching of the Er^{3+} $^2H_{11/2}$ and $^4S_{3/2}$ excited states due to energy transfer into a non-radiative

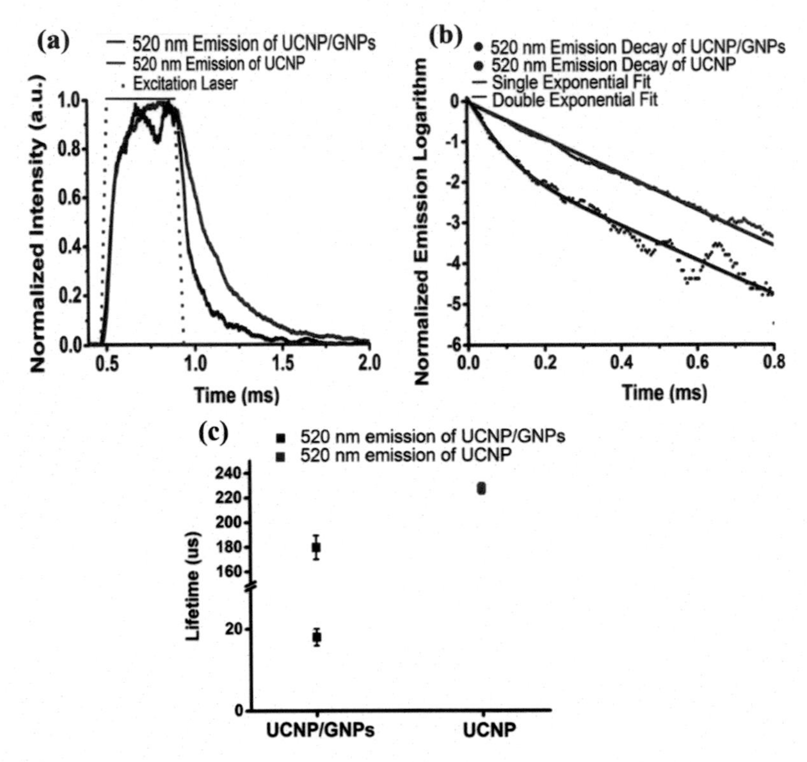

Fig. 7.5 **a** The excitation and decay profile of luminescence emission of plain UCNPs and UCNP/GNPs under 980 nm laser. b Single and double exponential decay of emission of bare and UCNP/GNPs under 980 nm laser, **c** the lifetime emission for 520 nm band for plain UCNPs and UCNP/GNPs under 980 nm laser. Reprinted with permission from *ACS Photonics, 2017, 4(7), pp 1864–1869*. Copyright 2017 American Chemical Society

Table 7.1 Lifetime and coefficient values for UCNP and UCNP/GNPs at room temperature

Fluorescent emitter	Band (nm)	Exponential coefficient	Lifetime (us)
UCNP	520	1.0	169.6 ± 4
	540	1.0	180 ± 1
		0.79 ± 0.02	65 ± 6
UCNP/GNPs	520	0.27 ± 0.02	276 ± 11
		0.76 ± 0.01	133 ± 9
	540	0.32 ± 0.07	327 ± 1

Reprinted with permission from *ACS Photonics, 2017, 4(7), pp 1864–1869*. Copyright 2017 American Chemical Society

energy level. We confirmed this hypothesis by measuring the temperature dependence in the steady-state emission from the H band for UCNPs shown in Fig. 7.6b. The steady-state emission intensity is normalized to the area under the H band at

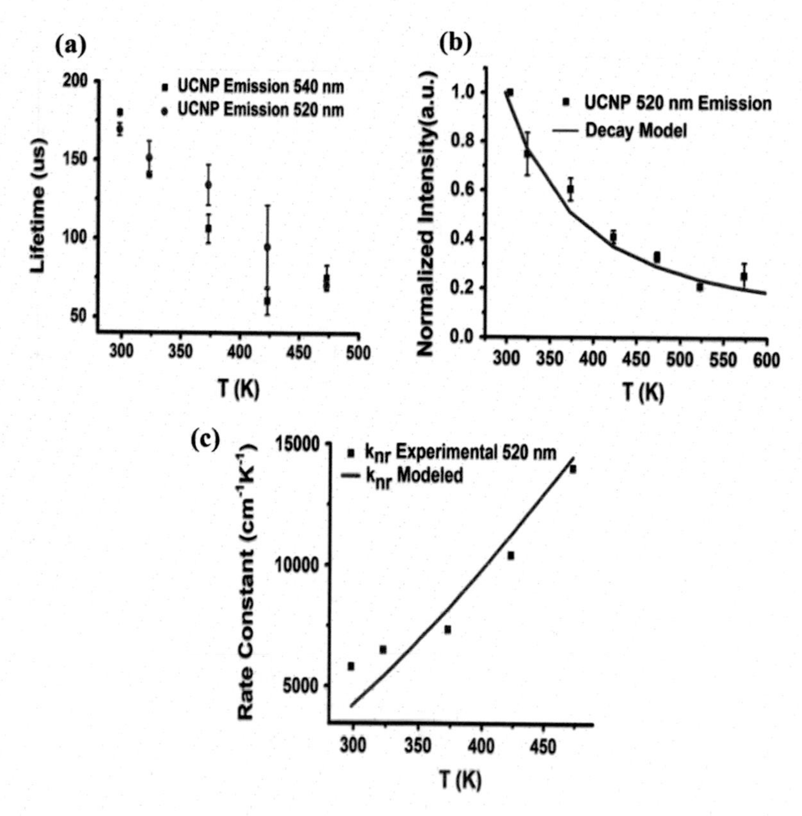

Fig. 7.6 **a** Lifetime drop for H and S bands as a function of temperature for UCNPs, **b** intensity drop of emission for H band as the temperature increases from 25 to 300 °C and, **c** non-radiative rate constant increase as a function of temperature for UCNPs under 980 nm excitation laser. Reprinted with permission from *ACS Photonics, 2017, 4(7), pp 1864–1869*. Copyright 2017 American Chemical Society

298 K. A nearly exponential drop in emission intensity is observed with temperature. We can model this behavior assuming two parallel pathways between a non-radiative and radiative path. The non-radiative path is given by a rate constant with a temperature dependence that can be expressed with an Arrhenius expression, $k_{nr} = A\ \exp(\frac{-E_a}{kT})$. The steady-state emission is proportional to the quantum yield (Φ) given by $\phi = \frac{k_r^o}{k_r^o + k_{nr}}$ where k_r^o is the intrinsic radiative rate constant (reciprocal of the intrinsic lifetime). We use an initial guess of the room temperature quantum yield of 0.5% for our relatively large UCNPs to set the intrinsic lifetime at 17 ms. Our model fit is not sensitive to the initial guess in quantum yield as long as a value less than 5% is used. In this model, quenching of the excited state increases with temperature because the non-radiative rate constant increases with temperature. The solid red line in Fig. 7.6b is the model fit to the temperature dependent drop in H band emission. The non-radiative rate constant can also be determined from the

observed lifetime for the UCNPs. In the two-pathway model presented above, the observed rate constant is the sum of the intrinsic radiative rate constant and the non-radiative rate constant, i.e., $k_{obs} = k_r^o + k_{nr}$. The observed lifetime is the reciprocal of the observed rate constant given in Fig. 7.6a. The non-radiative rate constant is then calculated from the observed and intrinsic rate constant.

The activation energy (E_a) for the model fit is 700 cm^{-1} with a pre-exponential factor (A) of 1.5 × 10^5. This low barrier value for the activation energy suggests that non-radiative relaxation is occurring by multi-phonon relaxation to lower Er^{3+} levels. We tested this hypothesis by measuring the visible spectrum of the green bands ($^2H_{11/2} \rightarrow {}^4I_{15/2}$ and $^4S_{3/2} \rightarrow {}^4I_{15/2}$ transitions) and red bands ($^2F_{5/2} \rightarrow {}^4I_{15/2}$) as temperature is increased. A two-dimensional correlation analysis is also performed on the temperature dependence in the emission spectrum for the spectral region 500–700 nm to see if the loss of green band emission is correlated with a gain in the red band emission ($^2F_{5/2} \rightarrow {}^4I_{15/2}$ transition). The emission spectra are shown in Fig. 7.7a while the correlation map is shown in Fig. 7.7b. A negative correlation is observed between the green bands and red bands confirming that loss in the H and S band is correlated with a gain in the red bands. The energy difference between the green bands and red bands is ~3500 cm^{-1}. If a single phonon has the energy of ~700 cm^{-1}, then it would take ~5 phonons to make the energy gap of 3500 cm^{-1}.

Because the emission bands used to calculate temperature are quenched when interacting with gold nanocrystals and are quenched with temperature, it is possible that the quenching rate is different between the H and S bands. To probe this temperature effect, we measured the temperature dependence in the steady-state emission from the H and S bands for UCNPs and UCNP/GNPs. The sample

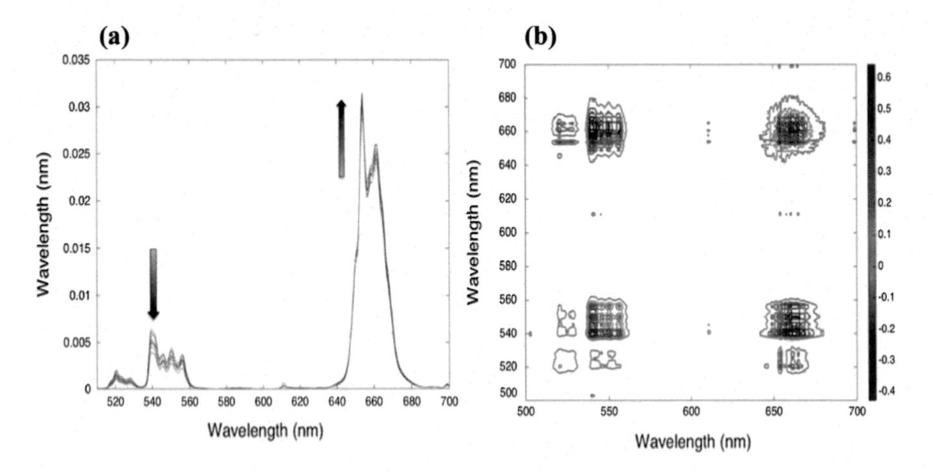

Fig. 7.7 **a** Normalized emission spectra taken over the temperature range of 10–90 °C. **b** Two-dimensional correlation map of spectrum shown in (**a**). Reprinted (adapted) with permission from *ACS Photonics, 2017, 4 (7), pp 1864–1869*. Copyright 2017 American Chemical Society

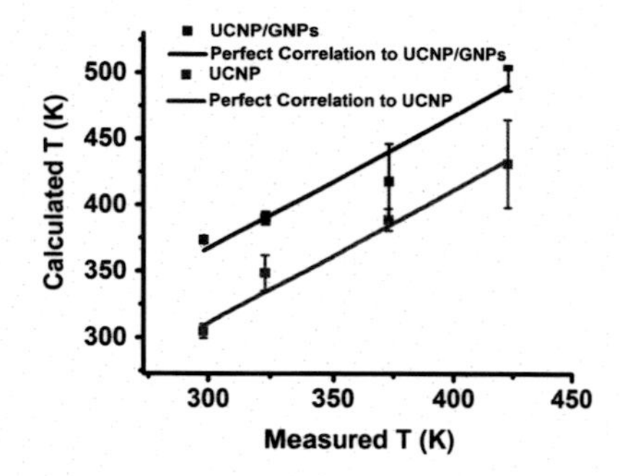

Fig. 7.8 Calculated temperature as a function of measured temperature for UCNP/GNPs and UCNPs under 980 nm excitation laser irradiation. Reprinted with permission from *ACS Photonics, 2017, 4(7), pp 1864–1869*. Copyright 2017 American Chemical Society

temperature was recorded using a thermocouple and the recorded temperature was compared with the temperature calculated using the steady-state emission of the H and S bands using the identical set-up we measured the lifetime data.

These results shown in Fig. 7.8 is plotted for the calculated temperature against the measured temperature. The calculated temperature for UCNPs and UCNP/GNPs is shown as red and black squares respectively. The uncertainty in the calculated temperature increases with temperature because the emission is quenched with temperature. The calculated temperature for the UCNP/GNPs has a temperature increase offset of ~60 K compared to the UCNPs because the UCNP/GNPs absorb more 980 nm light than the UCNPs. A linear relationship is observed, as expected, from this plot with unity slope and zero intercept. Higher temperature results were difficult to obtain because the signal is low, and the uncertainty is large due to quenching of the emission. Figure 7.8 shows a good correlation between calculated and measured temperature, even for the UCNPs with gold nanocrystals, confirming that the H and S band temperature dependent quenching rates are very similar and that the UCNPs with and without gold nanocrystals can be used as temperature sensors even at a temperature as high as 450 K.

References

1. Boyer JC, van Veggel F (2010) Absolute quantum yield measurements of colloidal NaYF4: Er^{3+}, Yb^{3+} upconverting nanoparticles. Nanoscale 2(8):1417–1419
2. Schietinger S, Aichele T, Wang H-Q, Nann T, Benson O (2010) Plasmon-enhanced upconversion in single NaYF4:Yb^{3+}/Er^{3+} codoped nanocrystals. Nano Lett 10(1):134–138
3. Debasu ML, Brites CDS, Balabhadra S, Oliveira H, Rocha J, Carlos LD (2016) Nanoplatforms for plasmon-induced heating and thermometry. ChemNanoMat 2(6):520–527
4. Rodriguez-Sevilla P, Zhang YH, Haro-Gonzalez P, Sanz-Rodriguez F, Jaque F, Sole JG, Liu XG, Jaque D (2016) Thermal scanning at the cellular level by an optically trapped upconverting fluorescent particle. Adv Mater 28(12):2421–2426

5. Rohani S, Quintanilla M, Tuccio S, De Angelis F, Cantelar E, Govorov AO, Razzari L, Vetrone F (2015) Enhanced luminescence, collective heating, and nanothermometry in an ensemble system composed of lanthanide-doped upconverting nanoparticles and gold nanorods. Adv Opt Mater 3(11):1606–1613
6. Debasu ML, Ananias D, Pastoriza-Santos I, Liz-Marzan LM, Rocha J, Carlos LD (2013) All-in-one optical heater-thermometer nanoplatform operative from 300 to 2000 K based on Er^{3+} emission and blackbody radiation. Adv Mater 25(35):4868–4874
7. Sedlmeier A, Achatz DE, Fischer LH, Gorris HH, Wolfbeis OS (2012) Photon upconverting nanoparticles for luminescent sensing of temperature. Nanoscale 4(22):7090–7096
8. Fischer LH, Harms GS, Wolfbeis OS (2011) Upconverting nanoparticles for nanoscale thermometry. Angew Chem Int Edit 50(20):4546–4551
9. Fujii M, Nakano T, Imakita K, Hayashi S (2013) Upconversion luminescence of Er and Yb codoped NaYF4 nanoparticles with metal shells. J Phys Chem C 117(2):1113–1120
10. Su Q, Han S, Xie X, Zhu H, Chen H, Chen C-K, Liu R-S, Chen X, Wang F, Liu X (2012) The effect of surface coating on energy migration-mediated upconversion. J Am Chem Soc 134 (51):20849–20857
11. Lu D, Cho SK, Ahn S, Brun L, Summers CJ, Park W (2014) Plasmon enhancement mechanism for the upconversion processes in NaYF4:Yb^{3+}, Er^{3+} nanoparticles: Maxwell versus Förster. ACS Nano 8(8):7780–7792
12. Boyer JC, Cuccia LA, Capobianco JA (2007) Synthesis of colloidal upconverting NaYF4: Er^{3+}/Yb^{3+} and Tm^{3+}/Yb^{3+} monodisperse nanocrystals. Nano Lett 7(3):847–852
13. Bogdan N, Vetrone F, Ozin GA, Capobianco JA (2011) Synthesis of ligand-free colloidally stable water dispersible brightly luminescent lanthanide-doped upconverting nanoparticles. Nano Lett 11(2):835–840

Chapter 8
Near Field Nanoscale Temperature Measurement Using AlGaN:Er^{3+} Film via Photoluminescence Nanothermometry

Susil Baral, Ali Rafiei Miandashti and Hugh H. Richardson

8.1 Introduction

Advancement in many photothermal technologies is predicated on a fundamental understanding of how scaling laws affect heat generation, heat dissipation and temperature distributions (both transient and steady-state) [1–8]. One basic question is how the thermal profile changes when comparing a thermal point source to a different sized heated structure such as a nanoparticle, virus, transistor, cell, tumor or bodily organ such as heart or kidneys? The temperature distribution from a thermal point source immersed in a medium of finite thermal conductivity falls off with inverse distance away from the point source. This behavior is observed in a theoretical thermal profile from a single 60 nm diameter gold nanoparticle immersed in water where the distance outside of the nanoparticle decays to half of the maximum temperature in 30 nm [9, 10]. However, an ensemble of gold nanorods inside a HeLa cell has a thermal profile outside of the cell that decays to a half temperature in 5 μm. The extent that temperature changes outside the optical heated region is important in photothermal therapy where destruction of healthy tissue should be minimized outside the tumor. One strategy to minimize damage to normal tissue is to use temperature-feedback to adjust the heat generation over time [11]. This strategy requires monitoring the temperature during heating. Unfortunately, temperature measurements become increasingly more difficult as the size of the system decreases especially if system size is below the diffraction limit of light.

The spatial resolution of conventional optical imaging is limited by the diffraction limit of light. Because of this limitation, traditional far field measurements fail to provide spatial details from objects smaller than the diffraction limit. Also, optical temperature measurements are distorted because the local temperature of a sub-diffraction hot object is a convolution of the real temperature with the point spread function of the microscope. Numerous techniques have been developed in recent years that offer sub-diffraction measurements, including the Nobel-prize

A. R. Miandashti et al., *Photo-Thermal Spectroscopy with Plasmonic and Rare-Earth Doped (Nano)Materials*, Nanoscience and Nanotechnology,
https://doi.org/10.1007/978-981-13-3591-4_8

winning work on super-resolution microscopy and imaging [12–14]. Scanning Near-field Optical Microscopy (SNOM) is a sub-diffraction imaging technique where the diffraction limit of the light is broken by scanning through a sharp tip in the near field of the object and collecting the evanescent wave information [15]. SNOM techniques yield sub-diffraction images where spatial topological information is ascertained but added information such as temperature topological maps is rare. Adding this extra dimension of sub-diffraction limited temperature/spatial information will make immediate contribution to many nanoscience and nanotechnology applications.

Our approach to measure the local temperature of an optically excited gold nanostructure is to combine near-field microscopy with Erbium ion photoluminescence. We show in this chapter that this approach yields thermal images where the spatial temperature resolution is limited by the SNOM cantilever tip aperture and is adequate to measure the true local temperature from optically excited nanostructures. The SNOM aperture tip can also be used collect the Erbium ion photoluminescence from a large illumination area where the steady-state thermal profile of an optically excited nanoparticle cluster is imaged. We use this new approach to measure the steady-state thermal profiles from different sized clusters of 40 nm diameter gold nanoparticles and show that the maximum temperature for the cluster scales with the size of the nanoparticle cluster. We measure the distance from the edge of a cluster to the point where the maximum temperature in the steady state thermal profile is reduced by a factor of 2 ($r_{1/2}$) and find that there is a linear relationship with cluster radius. Finally, we can use the collective heating from a two-dimensional array of nanoparticles to determine the average spacing between particles.

8.2 Characterization of NSOM Tip and Nanoparticles

The NSOM tip used in this method was characterized using scanning electron microscope. Figure 8.1 shows the SEM image of a pyramidal shape NSOM tip and the approximate size of aperture in our temperature measurements. The aperture is a rectangular shape with a size of $\sim 100 \times 120$ nm. The size of the hole becomes larger after a few times of usage and the resolution of the images can no longer be below the diffraction limit of light when the aperture size gets bigger therefore the tip needs to be replaced after it becomes blunt. Figure 8.1b shows a representative image of 40 nm gold nanoparticles that are used for photothermal heat generation in sub-diffraction temperature measurements.

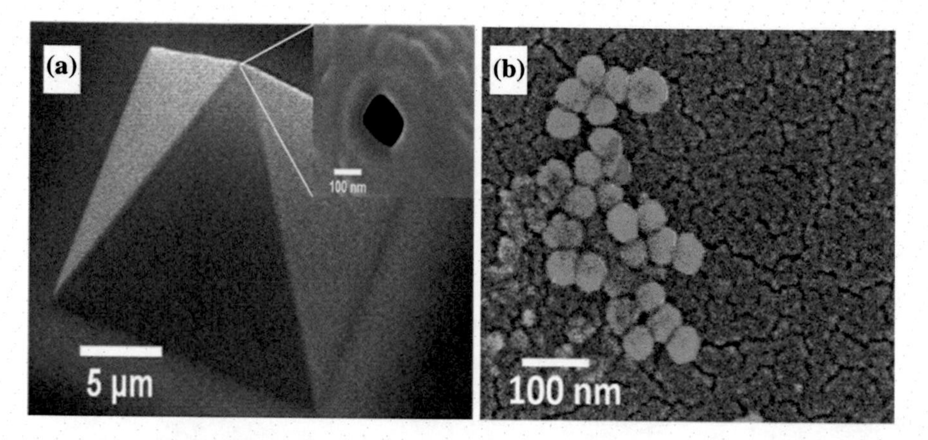

Fig. 8.1 **a** Characterization of NSOM tip and aperture size. NSOM tip has a pyramidal shape with a size of approximately 15 × 15 um base. Inset shows an aperture size of 100 nm by 120 nm rectangular hole at the apex of the pyramid. **b** Representative SEM image of a cluster of 40 nm gold nanoparticles drop-casted onto the Silicon substrate. The substrate was made hydrophilic by plasma treatment before drop-casting the dilute colloidal solution of gold nanoparticles

8.3 Sub Diffraction Near Field Photothermal Temperature Measurement

Characterization of near-field thermal measurement technique with topside illumination is shown in Fig. 8.2. A schematic (Fig. 8.2a) for the optical setup with 532 nm CW laser excitation through the SNOM tip shows how the two lasers work together to excite (532 nm laser) and collect Er^{3+} emission from a small area limited by the tip aperture and the 980 nm laser is used as feedback for the tip position. Nanostructures (lithographically fabricated nanostructures or drop-casted gold nanoparticles) on top of the thermal sensor film [16] of $Al_{0.94}$ $Ga_{0.06}$ N embedded with Er^{3+} ions (~300 nm thick) on a silicon substrate are optically excited with 532 nm CW laser through the SNOM cantilever tip (tip aperture ~120 nm) and the emission from the thermally coupled Er^{3+} energy levels is back collected through the tip mounted on 50X ultra long working distance objective (50X, NA 0.55) coupled with CCD spectrograph via 100 µm optical fiber. Data collection and optical measurements are performed in a near-field contact mode of a Scanning Near-Field Optical Microscope (WITec α-SNOM300 s). Figure 8.2b shows the representative Er^{3+} photoluminescence spectra at a hot spot and at a background position on the substrate. The Er^{3+} ions are excited with 532 nm laser light resulting in a typical Er^{3+} photoluminescence spectrum (see Fig. 8.2b). The relative intensities of the $^2H_{11/2} \rightarrow {}^4I_{15/2}$ and the $^4S_{3/2} \rightarrow {}^4I_{15/2}$ energy transitions of the Er^{3+} ions are temperature dependent [17, 18]. The relative intensities are related by a Boltzmann factor $\left(\frac{I_1}{I_2} = \left(\frac{g_1}{g_2}\right)\exp\left(-\frac{\Delta E}{kT}\right)\right)$ where (g_1/g_2) is the degeneracy factor for the upper state ($^2H_{11/2}$) compared to the lower state ($^4S_{3/2}$), ΔE is the energy difference between the two levels, k is the Boltzmann constant, and T is the absolute temperature. At higher temperatures (the region with optically heated nanostructures), the emission intensity of the high energy Er^{3+} emission band ($^2H_{11/2} \rightarrow {}^4I_{15/2}$) increases with respect to that of the low energy band ($^4S_{3/2} \rightarrow {}^4I_{15/2}$). Energy diagram of Er^{3+} and the corresponding temperature dependent photoluminescence from Er^{3+} excited states to the ground state is presented in previous chapters. Thermal images are constructed by integrating the Er^{3+} emission in each pixel location of the emission image, taking the ratio of the intensities of high energy to the low energy band (I_1/I_2) and applying the Boltzmann factor above. The degeneracy of the two state is determined by adjusting the degeneracy factor to give the extrapolated zero laser temperature to room temperature.

Figure 8.2c shows the temperature-time spectrum of optically heated gold nanoparticles spin-coated onto the thermal sensor film collected through the SNOM cantilever tip. An optically heated hot spot is first located by performing a survey image scan. Once the position of the hot spot is located, a time spectrum is taken of

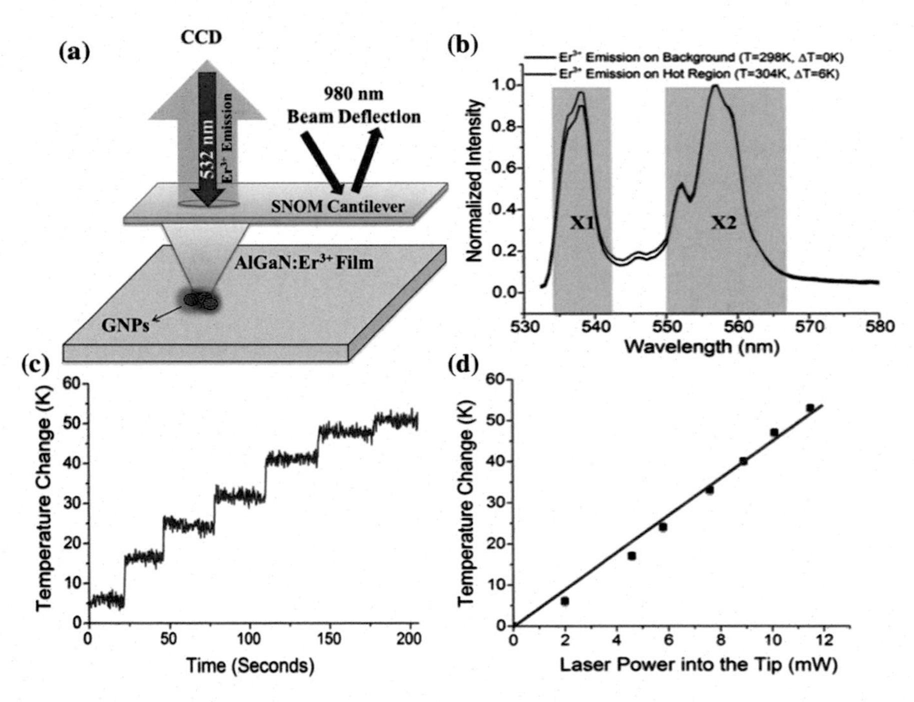

Fig. 8.2 **a** Schematic optical setup for the near-field thermal measurement technique with 532 nm CW laser excitation through the SNOM tip. **b** Representative Er^{3+} photoluminescence spectra on various regions on the substrate. **c** Temperature-time spectrum of the optically heated gold nanoparticles spin-coated onto the sapphire substrate with thermal sensor film of AlGaN:Er^{3+} collected through the SNOM cantilever tip. **d** Plot of a local temperature change (K) as a function of the laser power into the tip (mW) for temperature-time spectrum shown in c. Reprinted with permission from *Nanoscale, 2017, 4(7), pp 1864–1869*. Copyright 2017 Royal Society of Chemistry

the optically heated nanoparticles while the laser power is increased stepwise. The corresponding temperature change at each laser power is determined continuously from the Er^{3+} emission collected. In the temperature-time spectrum, the uncertainty in the temperature measurement is ~ 1 K with an integration time of 0.2 s. This gives a temperature accuracy of 0.45 K $Hz^{-1/2}$. A plot of a local temperature change (ΔT) as a function of the laser power into the tip is shown in Fig. 8.2d and is generated from the temperature time spectrum shown in Fig. 8.2c. The plot shows a nearly linear relationship (small deviation at low laser power) between the local temperature change and the laser power.

The spatial resolution of near-field measurement technique is probed by scanning a 40 nm gold nanoparticle drop casted onto coverslip and AlGaN:Er^{3+} film on a sapphire coverslip. The image shown in Fig. 8.3a is produced from collecting the transmission of 532 nm CW laser (10 mW power) from the top through the SNOM cantilever. This figure shows the transmission image from the 532 nm laser from a small ensemble of 40 nm gold nanoparticle. Figure 8.3b shows the corresponding

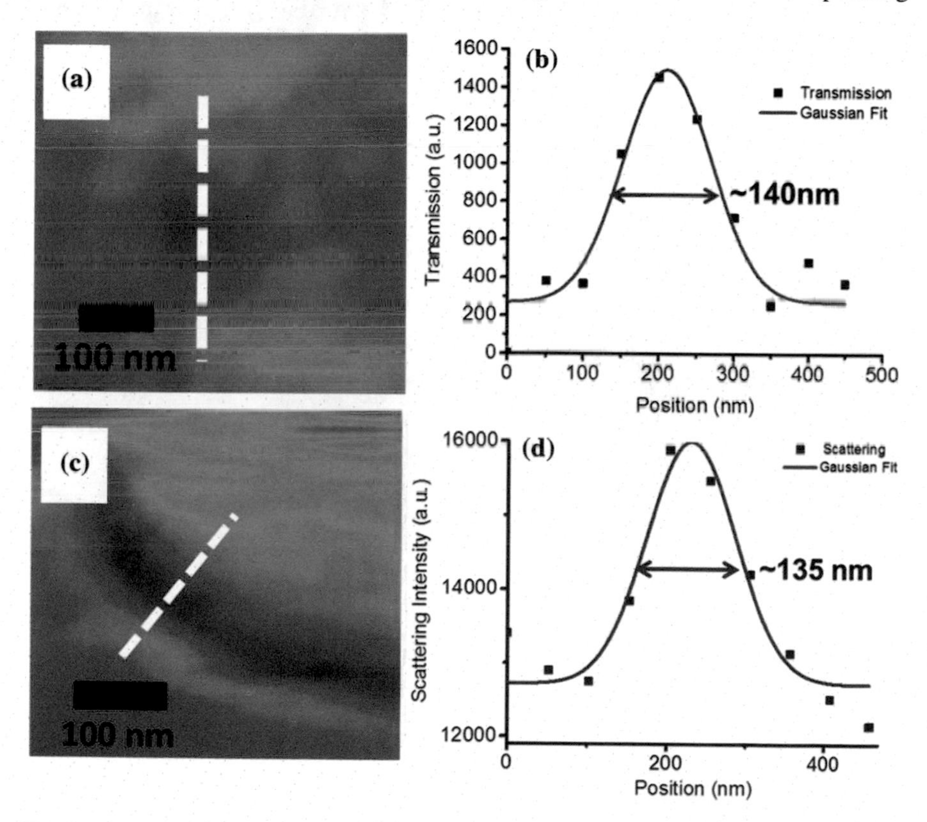

Fig. 8.3 Transmission and scattering images (**a**) and (**c**) and the corresponding transmission and scattering profiles (**b**) and (**d**) of the gold nanoparticles respectively. The transmission/scattering profiles show the spatial resolution of less than 150 nm for the dashed line drawn through the nanoparticles. Reprinted with permission from *Nanoscale, 2017, 4(7), pp 1864–1869*. Copyright 2017 Royal Society of Chemistry

532 nm inverse of transmission profile. The spatial resolution of ~140 nm is obtained as the Full Width at Half Maximum (FWHM) of a Gaussian function fitting to the scattering profile. Figure 8.3c shows the scattering image of a line of gold nanoparticles generated by drop-casting 40 nm gold nanoparticles onto the AlGaN:Er^{3+} film on a sapphire substrate. Figure 8.3d shows the corresponding scattering profile for the cross section (shown as dashed white line) drawn through the hot line on the near-field thermal image. The 532 nm scattering intensity profile is also fitted with Gaussian function with FWHM of ~135 nm. Importantly, in this image because the nanostructures and the film are not under constant illumination of 532 nm CW laser, and the laser is moving across the film along with the tip, this should not be considered as a steady state measurement.

Figure 8.4a shows the thermal image of a lithographically fabricated gold nanodot and nanosnake under near-field thermal measurement. The nanodot (200 nm diameter, 70 nm high) is excited through the tip with 532 nm CW laser (5 mW power into the tip). The near-field thermal image shows the hot-spot comparable to the true size of the nanodot. A SEM and AFM image of the nanodot

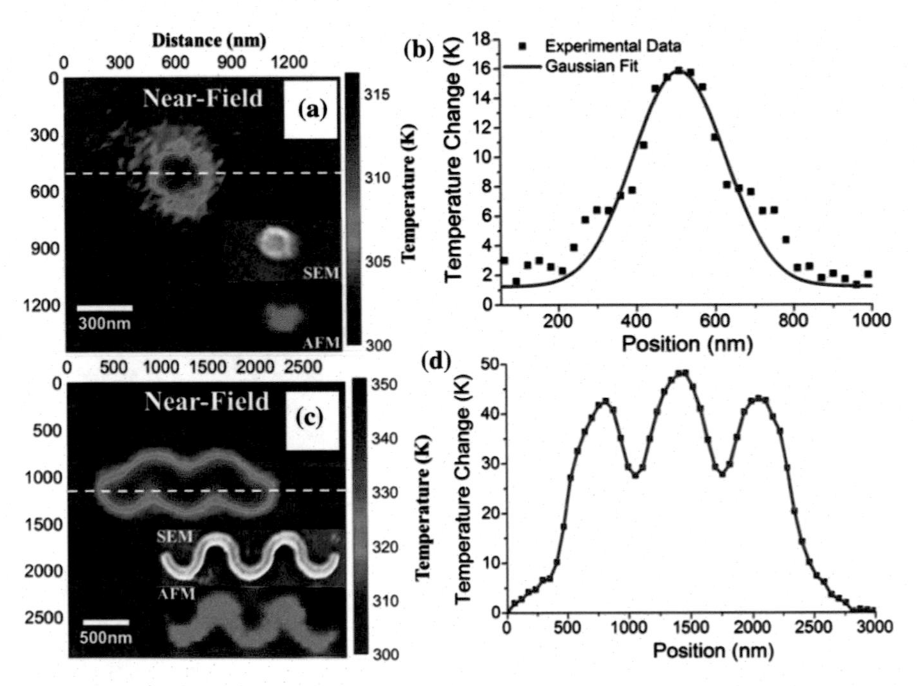

Fig. 8.4 Thermal images and the corresponding thermal profiles of lithographically fabricated gold nanodot and gold nanosnake under near-field thermal imaging measurements with 532 nm CW laser excitation through the SNOM tip. The near field thermal images show the hot-spot comparable to the true size of the nanostructures (For comparison, SEM and AFM images of the nanodot and nanosnake with same scale bar are shown as inset inside the near-field thermal images). Reprinted with permission from *Nanoscale, 2017, 4(7), pp 1864–1869*. Copyright 2017 Royal Society of Chemistry

with same scale are shown in the inset. Figure 8.4b shows the thermal profile drawn through the hot spot (white dotted line) of the thermal image shown in a. The FWHM of a Gaussian fit to the thermal profile is 230 nm. Figure 8.4c shows the thermal image of a lithographically fabricated gold nanosnake under near-field thermal measurement with 532 nm CW laser (7.5 mW power into the tip) excitation through the SNOM tip. The near-field thermal image is comparable to the true size of the nanosnake. The SEM and AFM image of the nanosnake with same scale are shown in the inset. Again, Fig. 8.4d is the thermal profile for a horizontal cross section (white dotted line) drawn through thermal image shown in c.

8.4 Steady State Near Field Photothermal Heat Study

Figure 8.5a shows the schematic optical setup for the steady-state temperature measurement under near field technique with 532 nm CW excitation from the bottom using a large laser spot (FWHM $\sim$10 μm). The Er^{3+} emission is collection in transmission mode through the SNOM tip. Nanostructures on top of the thermal sensor film of transparent AlGaN:Er^{3+} on sapphire are optically excited with 532 nm CW laser from the bottom objective (20X, N.A. 0.2) with the focused spot size of FWHM $\sim$10 μm. The laser simultaneously excites the Er^{3+} and heats the gold nanoparticles/nanostructures under the area of illumination. The emission from the thermally coupled Er^{3+} energy levels is then collected in transmission mode through the SNOM cantilever tip (tip aperture $\sim$100 nm) mounted on a 50X ultra long working distance objective (50x, NA 0.55) coupled with CCD spectrograph via 100 μm optical fiber. Figure 8.5b shows the intensity profile of a focused laser spot from the 532 nm CW laser. The FWHM of the spot size is approximately 10 μm and has a broad constant intensity region of about 6 μm. The large constant intensity excitation area compared to the small sampled area within the constant intensity region insures that steady state thermal images are collected with the SNOM tip. Figure 8.5c shows the thermal image of a gold nanoparticle cluster created by drop-casting 40 nm diameter nanoparticles onto the thermal sensor film. The image is collected under constant laser excitation of 40 mW with a laser spot diameter of 10 μm. This gives a laser intensity of 5×10^8 W/m^2. Figure 8.5d shows the thermal profiles for the cross section shown for the dotted white line in Fig. 8.5c. On this profile we have labeled ΔT_{max}, R_c, $r_{½}$, and a 1/r profile given by the equation, $\Delta T(r) = \frac{\Delta T_{max} R_c}{R+r}$ for $r \geq R_c$. In this profile r is a radial coordinate that starts at the edge of the cluster and proceeds outward. ΔT_{max} is the maximum temperature change, R_c is the cluster radius and $r_{½}$ is the length from the edge of the cluster to $\Delta T_{max}/2$. The diameter of the cluster is determined by measuring the length where ΔT_{max} is relatively constant and from one inflection point of the profile to the other.

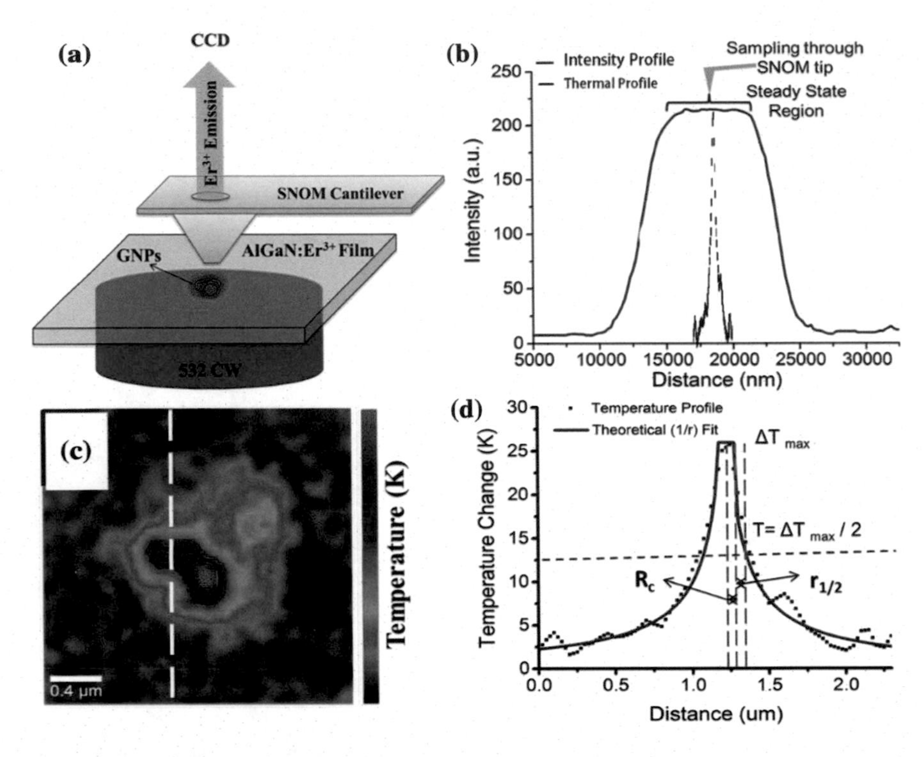

Fig. 8.5 **a** Schematic optical setup for the steady-state temperature measurement under near field technique. **b** Focused spot intensity profile of a bottom 532 nm CW laser. The FWHM of the spot size is ~10 μm and has a broad flat region of ~6 μm with constant intensity. **c** Thermal image of the gold nanoparticles drop-casted onto the thermal sensor film of AlGaN:Er^{3+} on sapphire glass substrate under steady state measurement in Near-Field imaging mode. **d** Thermal profile for the horizontal cross section (represented as dotted white line) drawn across the near-field thermal image shown in (**c**). Reprinted with permission from *Nanoscale, 2017, 4(7), pp 1864–1869.* Copyright 2017 Royal Society of Chemistry

8.4.1 Estimation of Cluster Radius (R_c) from Thermal Profile

The diameter of a cluster of nanoparticles is estimated from the experimental thermal profile generated by drawing a cross section across the thermal image as represented in Fig. 8.6d. On this thermal profile, diameter of a cluster is the region (length) where ΔT_{max} is relatively constant and from one inflection point of the profile to the other. The radius (R_c) of a cluster is then half this length (diameter). Some experimental thermal profiles and their estimated radius are shown in Fig. 8.6. In this figure thermal profile of several clusters with different sizes are plotted.

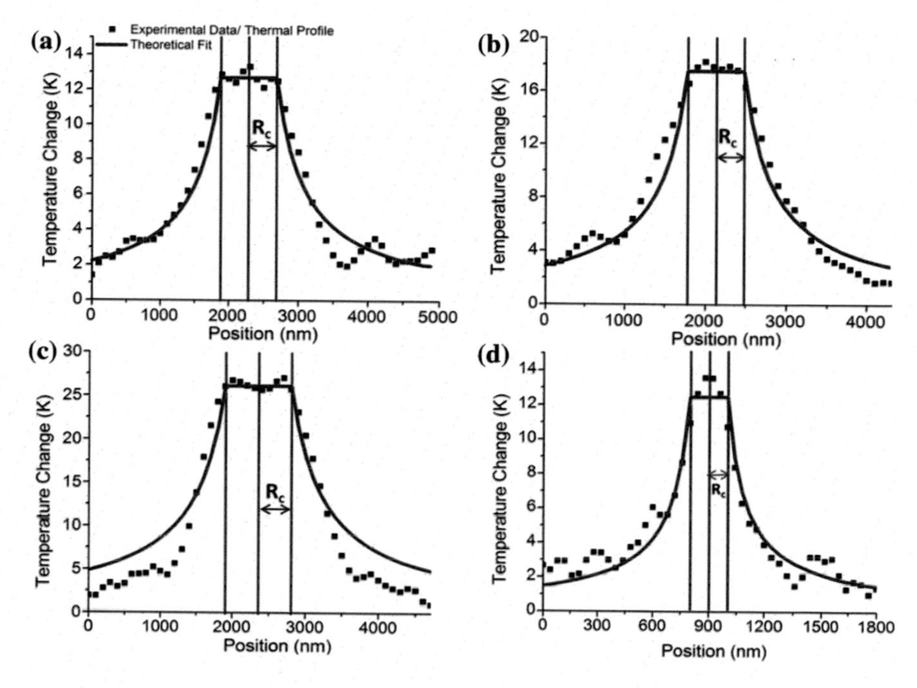

Fig. 8.6 Illustration of estimation of cluster radius (R_c) from thermal profile generated by drawing a cross section across the steady state thermal image of a cluster of 40 nm gold nanoparticles. Reprinted (adapted) with permission from *Nanoscale, 2017, 4(7), pp 1864–1869*. Copyright 2017 Royal Society of Chemistry

8.5 Comparison Between Estimation of Cluster Radius (R_c) from Thermal Profile, AFM, and Changes on Er^{3+} Luminescence Intensity

Figure 8.7 shows the comparison between estimation of cluster radius from (a) thermal profile (b) AFM profile and (c) total Er^{3+} photoluminescence intensity. The AFM profile (Fig. 8.7b) is generated by drawing a cross section across the AFM image of a cluster generated during near field imaging. The AFM profile shows the height of ~33 nm suggesting the 2-D cluster of the nanoparticles (40 nm diameter). The FWHM of the AFM profile is approximately 780 nm which is very close to the cluster diameter of 800 nm estimated from the thermal profile shown in Fig. 8.7a. The Er^{3+} intensity profile is generated by drawing a cross section across the Er^{3+} intensity image. The Er^{3+} intensity image is generated by applying a filter for Er^{3+} emission peaks (H and S band) on an original image/spectrum. As the measurement is performed by bringing the 532 nm CW laser from the bottom and collecting the Er^{3+} emission from the film on transmission mode, the region with gold nanoparticles cluster shields the Er^{3+} emission. The Er^{3+} intensity

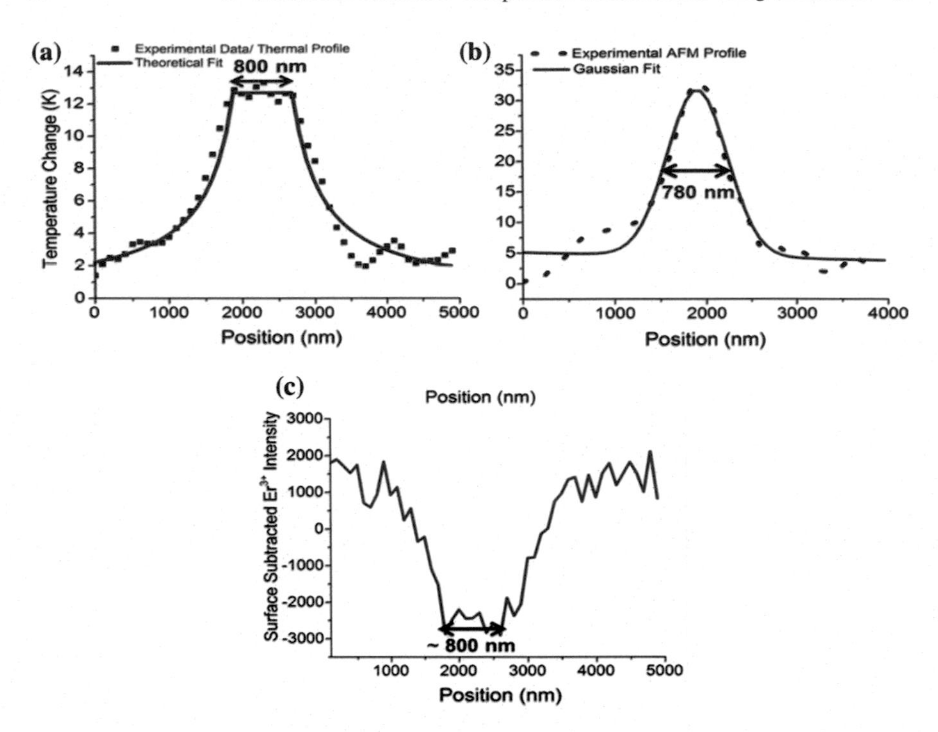

Fig. 8.7 **a** Steady state thermal profile of a cluster of gold nanoparticles drop-casted onto the thermal sensor film of AlGaN:Er^{3+} on sapphire glass substrate in Near-Field imaging mode. **b** Illustration of estimation of cluster radius (R_c) from AFM profile generated during the raster scanning of nanocluster. R_c is the radius of the cluster and is conceived from the thermal profile, AFM profile and total Er^{3+} intensity drop in (**c**). Reprinted (adapted) with permission from *Nanoscale, 2017, 4(7), pp 1864–1869*. Copyright 2017 Royal Society of Chemistry

profile (Fig. 8.7c) shows the drop in Er^{3+} intensity with the plateau of approximately 800 nm, in close agreement with the cluster diameter of 800 nm estimated from the thermal profile shown in a.

8.6 Two Laser Steady State Data Collection Experiment

Figure 8.8 shows the thermal image of drop casted gold nanoclusters under near-field thermal measurement. The cluster is excited through the tip with 532 nm CW laser (4.5 mW power). The near-field thermal image calculated from luminescence ratio thermometry shows the hot-spot in the center and cooler areas around the cluster. Figure 8.8b also shows the thermal image of the same cluster excited by 532 nm CW laser (4.5 mW power) through the tip and 980 nm CW laser from bottom (50 mW). Figure 8.8c shows the thermal image produced from subtraction of the thermal images in a and b. The resulting temperature profiles

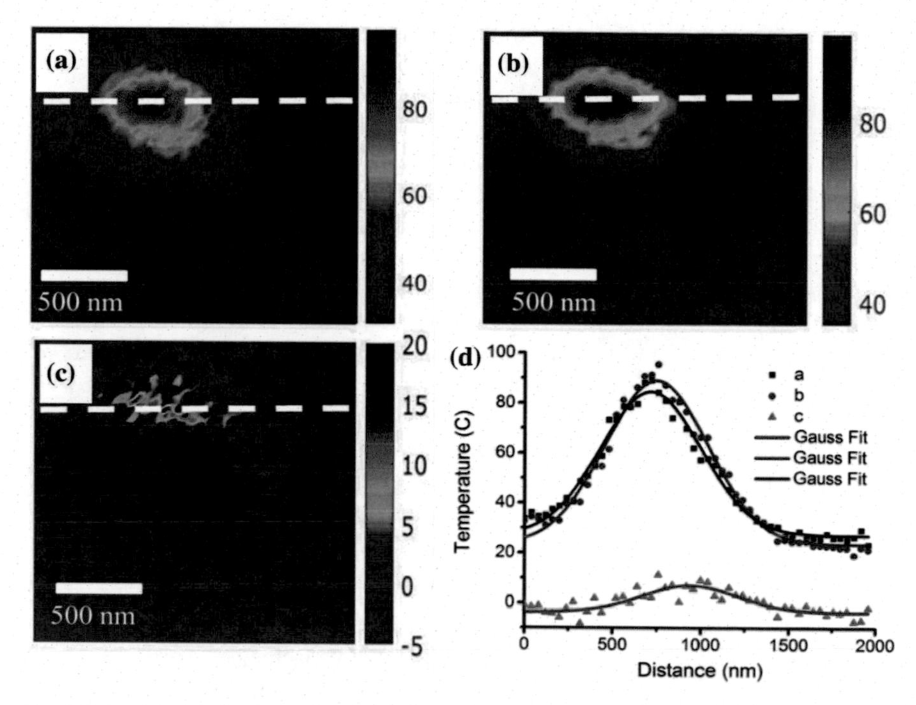

Fig. 8.8 Two laser temperature profile data collection from 40 nm drop casted gold nanoparticles, **a** Scanning near-field optical image of a cluster of gold nanoparticles illuminated by 532 nm CW as excitation laser through the SNOM tip. **b** Scanning near-field optical image of cluster of gold nanoparticles illuminated by 532 nm CW as excitation laser through the SNOM tip and 980 nm as heater from bottom. **c** Thermal image produced from subtraction of b from cluster of 40 nm gold nanoparticles. **d** Thermal profile from white dashed lines drawn across the cluster from images of illumination under 532 nm laser, 532 and 980 nm lasers and subtracted temperature image. Reprinted (adapted) with permission from *Nanoscale, 2017, 4(7), pp 1864–1869*. Copyright 2017 Royal Society of Chemistry

comparing with and without the second excitation laser for the horizontal white dashed line drawn across the hot spot is shown in Fig. 8.8d.

8.7 Scaling Law in Near Field Photothermal Heat Dissipation

Near-field microscopy is a sub-diffraction imaging technique where the diffraction limit of the light is broken by focusing the light through an aperture with a diameter less than the wavelength of light and collecting the evanescent wave from an optically excited sample. The maximum diffraction limited spatial resolution (d) that can be achieved under far-field measurements is $d \approx \frac{\lambda}{2NA}$, where λ is the wavelength of the light used and NA is the numerical aperture of the collection objective. For our measurements, the diffraction limit calculated using the above relation is ~490 nm (λ = 532 nm, NA = 0.55). A higher numerical aperture lens (such as 1.0) can reduce this diffraction limited resolution to ~270 nm. We have previously shown that for our microscope, the spatial resolution is dependent on the bore size of the optical collection fiber to the CCD camera [19] and the numerical aperture of our objective. Smaller bore collection fiber reduces both the light transmission through the fiber and the spatial resolution. In a near-field measurement, the spatial resolution is limited by the size of the tip aperture [20]. In our measurements, we excite (or collect emission from) the thermal sensor film through a near-field tip with aperture diameter of approximately 100 nm. The thermal spatial resolution of our measurement is expected to be around 100 nm. Clusters of nanoparticles with size greater than 100 nm can be thermally imaged with spatial resolution comparable to the true size of the nanostructures because the tip is in contact with the sample and only the region under the tip is sampled.

For a nanometer sized hot spot where the heat dissipation can be characterized with spherical symmetry, the temperature change due to optical excitation is given by $\Delta T_{NP} = \frac{C_{abs} I}{4\pi R_{eff} \beta k_{eff}}$ where C_{abs} is the absorption cross section, I is the laser intensity, k_{eff} is the effective thermal conductivity given by $k_{eff} = \frac{k_{sub} + k_{air}}{2} \approx \frac{k_{sub}}{2} = 0.75\mathrm{W/m\text{-}K}$, R_{eff} is the effective radius for a particle with equal volume to that of a sphere and β (thermal-capacitance coefficient) is a parameter that depends on the aspect ratio of the nanostructure [21]. For a spherical structure the aspect ratio and β are one. Non-spherical nanostructures have a thermal-capacitance coefficient greater than one because of increase heat dissipation into the surrounding due to increase surface to volume ratio compared to a sphere. The volume of a nanodot (approximated as a cylinder with volume $V = \pi r^2 l$) is used to calculate the effective radius $R_{eff} = \left(\frac{3V}{4\pi}\right)^{1/3}$. The absorption cross section for the lithographically prepared nanodot shown in Fig. 8.4A is $C_{abs} = 3.8 \times 10^{-14}\ \mathrm{m}^2$ [16]. The laser intensity absorbed by the nanodot for a temperature change of 15 K (see Fig. 8.4b) is $3 \times 10^8\ \mathrm{W/m}^2$ using the equation $I = \frac{\Delta T_{\max} 4\pi R_{eff} \beta k_{eff}}{C_{abs}}$ with an aspect ratio for the nanodot of 2.86 and a β value of 1.1. This intensity gives a laser power through the tip of ~4 µW if the laser spot size is the area of a circle with a tip radius of 60 nm.

The steady-state temperature measurements shown in Fig. 8.5 are made on a number of different sized clusters. The collective heating from a two-dimensional ensemble of gold nanoparticles with spacing between particle centers (Δ) is given by $\Delta T_{tot}(r) \approx \Delta T^{o}_{\max} \frac{R_{Au}}{\Delta} N_{NP}^{1/2}$ [9]. In this equation, $\Delta T^{o}_{\max}$ is the maximum temperature change for a single gold nanoparticle with radius R_{Au}, and N_{NP} is the number of nanoparticles in the cluster given by $\rho_C \left(\frac{R_C}{R_{Au}}\right)^2$ where R_C is the cluster radius and ρ_c is the ratio of the total two-dimensional area occupied by the nanoparticles to the total area of the nanoparticle cluster [22, 23]. This value changes with R_C but converges to $\frac{\pi}{2\sqrt{3}} = 0.9069$ for an infinite sized container (cluster) with hexagonal packing. An average value for our clusters sizes with hexagonal close packing is ~0.85. Substituting this for N_{NP} and dividing by the laser intensity gives $\frac{\Delta T_{tot}}{I} = \frac{\Delta T^{o}_{max} R_C \rho_C^{1/2}}{I\Delta}$. Our two-dimensional clusters can be approximated as two-dimensional disks with height d and radius R_c. The maximum change in temperature in the disk can be approximated as $\frac{\Delta T_{tot}}{I} = \frac{\Delta T^{o}_{max} R_C \rho_C^{1/2}}{I\Delta\beta}$ where β is a parameter that depends on the aspect ratio of the disk and takes into account more efficient heat release from a disk compared to a sphere [21]. This equation combines collective heating from a two-dimensional ensemble of gold nanoparticles [9] with temperature from an arbitrary shape such as a disk [21]. The β parameter is calculated using $\beta = \exp(0.040 - 0.0124\ln(x) + 0.0677\ln^2(x) - 0.00457\ln^3(x))$ where x is $2R_c/d$. Adding β to the collective heating for a two-dimensional ensemble gives a total temperature change for the two-dimensional ensemble that is expected to increase with R_c.

Figure 8.9A plots $\Delta T_{tot}/I$ with R_c where the total temperature change increases with R_c as expected. The slight bend in the curve is caused from changes in β because it depends upon aspect ratio that changes with cluster size. The solid red line is a theoretical fit to the data using $\frac{\Delta T_{tot}}{I} = \frac{\Delta T^{o}_{max} R_C \rho_C^{1/2}}{I\Delta\beta}$ and using $\frac{\rho_C^{1/2}}{\Delta}$ as the fitting parameter. The best fit yields a value for Δ of 43 nm when ρ_c is 0.85. This value is reasonable consistent with a two-dimensional film of closed-pack particles where the average spacing should be 40 nm.

Figure 8.9b shows that there is a linear relationship between $r_{1/2}$ and cluster radius. The temperature profile far from the nanostructure should fall off as 1/r where r is the distance that starts at the edge of the nanostructure. The expected

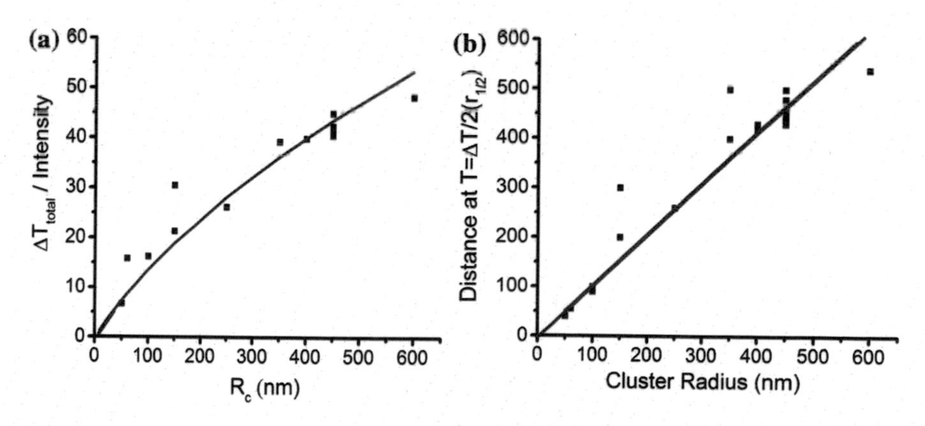

Fig. 8.9 **a** Temperature change under optical excitation normalized by the excitation laser intensity (ΔT_{total}/Intensity) plotted as a function of a cluster radius (R_c) normalized by β parameter for a steady state measurement on a number of different sized clusters. **b** A plot of a distance at which the temperature goes to half the maximum temperature change, $T = \Delta T_{max}/2$ represented as $r_{1/2}$ on a thermal profile shown in Fig. 8.5d versus the radius of a cluster (R) for a steady state measurement on a number of different sized clusters. Reprinted (adapted) with permission from *Nanoscale, 2017, 4(7), pp 1864–1869*. Copyright 2017 Royal Society of Chemistry

distance value at $\frac{\Delta T_{max}}{2}$ $(r_{1/2})$ is R_c. A plot of $r_{1/2}$ versus R_c should be linear with a slope of one and an intercept that goes through the origin. The slope is 1.03 ± 0.16 with a y intercept set to 0.

Figure 8.9 shows how scaling of the thermal profile into a surrounding media is affected by the size of the nanostructure optically heated. For a single 40 nm diameter gold nanoparticle, the heating of the nanoparticle produces a local temperature profile outside of the nanoparticle that dissipates to half temperature within ~20 nm. If, on the other hand, a large object, such as a 10 μm diameter cell, is filled with nanoparticles with an average distance that produces collective heating, then the temperature profile outside of the cell decays to a half temperature around 5 μm. Generally, if collective heating is occurring, then the maximum temperature change and temperature decay into the surrounding medium increases with the size of the heated object.

References

1. Fang Z, Zhen Y-R, Neumann O, Polman A, Garcia Javier, de Abajo F, Nordlander P, Halas NJ (2013) Evolution of light-induced vapor generation at a liquid-immersed metallic nanoparticle. Nano Lett 13(4):1736–1742
2. Polman A (2013) Solar steam nanobubbles. ACS Nano 7(1):15–18
3. El-Sayed IH, Huang XH, El-Sayed MA (2006) Selective laser photo-thermal therapy of epithelial carcinoma using anti-EGFR antibody conjugated gold nanoparticles. Cancer Lett 239(1):129–135
4. Hirsch LR, Stafford RJ, Bankson JA, Sershen SR, Rivera B, Price RE, Hazle JD, Halas NJ, West JL (2003) Nanoshell-mediated near-infrared thermal therapy of tumors under magnetic resonance guidance. Proc Natl Acad Sci USA 100(23):13549–13554
5. Huang XH, Jain PK, El-Sayed IH, El-Sayed MA (2008) Plasmonic photothermal therapy (PPTT) using gold nanoparticles. Lasers Med Sci 23(3):217–228
6. Park JH, von Maltzahn G, Ong LL, Centrone A, Hatton TA, Ruoslahti E, Bhatia SN, Sailor MJ (2010) Cooperative nanoparticles for tumor detection and photothermally triggered drug delivery. Adv Mater 22(8):880–885
7. Alkilany AM, Thompson LB, Boulos SP, Sisco PN, Murphy CJ (2012) Gold nanorods: their potential for photothermal therapeutics and drug delivery, tempered by the complexity of their biological interactions. Adv Drug Deliv Rev 64(2):190–199
8. Alvarez-Puebla RA, Liz-Marzan LM (2012) Traps and cages for universal SERS detection. Chem Soc Rev 41(1):43–51
9. Govorov AO, Zhang W, Skeini T, Richardson H, Lee J, Kotov NA (2006) Gold nanoparticle ensembles as heaters and actuators: melting and collective plasmon resonances. Nanoscale Res Lett 1(1):84–90
10. Govorov AO, Richardson HH (2007) Generating heat with metal nanoparticles. Nano Today 2(1):30–38
11. Rodriguez-Sevilla P, Zhang YH, Haro-Gonzalez P, Sanz-Rodriguez F, Jaque F, Sole JG, Liu XG, Jaque D (2016) Thermal scanning at the cellular level by an optically trapped upconverting fluorescent particle. Adv Mater 28(12):2421–2426
12. Betzig E, Patterson GH, Sougrat R, Lindwasser OW, Olenych S, Bonifacino JS, Davidson MW, Lippincott-Schwartz J, Hess HF (2006) Imaging intracellular fluorescent proteins at nanometer resolution. Science 313(5793):1642–1645

13. Hell SW, Wichmann J (1994) Breaking the diffraction resolution limit by stimulated-emission —stimulated-emission-depletion fluorescence microscopy. Opt Lett 19(11):780–782
14. Moerner WE, Fromm DP (2003) Methods of single-molecule fluorescence spectroscopy and microscopy. Rev Sci Instrum 74(8):3597–3619
15. Xiong Y, Liu Z, Sun C, Zhang X (2007) Two-dimensional Imaging by far-field superlens at visible wavelengths. Nano Lett 7(11):3360–3365
16. Carlson MT, Khan A, Richardson HH (2011) Local temperature determination of optically excited nanoparticles and nanodots. Nano Lett 11(3):1061–1069
17. Gurumurugan K, Chen H, Harp GR, Jadwisienczak WM, Lozykowski HJ (1999) Visible cathodoluminescence of Er-doped amorphous AlN thin films. Appl Phys Lett 74(20):3008–3010
18. Garter MJ, Steckl AJ (2002) Temperature behavior of visible and infrared electroluminescent devices fabricated on erbium-doped GaN. IEEE Trans Elect Dev 49(1):48–54
19. Baral S, Johnson SC, Alaulamie AA, Richardson HH (2016) Nanothermometry using optically trapped erbium oxide nanoparticle. Appl Phys Mater Sci Process 122:(4)
20. Hecht B, Sick B, Wild UP, Deckert V, Zenobi R, Martin OJF, Pohl DW (2000) Scanning near-field optical microscopy with aperture probes: fundamentals and applications. J Chem Phys 112(18):7761–7774
21. Baffou G, Quidant R, de Abajo FJG (2010) Nanoscale control of optical heating in complex plasmonic systems. ACS Nano 4(2):709–716
22. Specht E The best known packings of equal circles in a circle (complete up to N = 2600). http://hydra.nat.uni-magdeburg.de/packing/cci/cci.html
23. Graham RL, Lubachevsky BD, Nurmela KJ, Ostergard PRJ (1998) Dense packings of congruent circles in a circle. Discret Math 181(1–3):139–154

Printed in the United States
By Bookmasters